To Sandi ~

With love,

June Scobee Rodgers

D0425844

Silver Linings

Triumph of the Challenger 7

"*Nothing* happens unless
first a dream."

– *Carl Sandburg*

Silver Linings

Triumph of the *Challenger* 7

June Scobee Rodgers
wife of the commander of *Challenger*

PEAKE ROAD ®

Macon, Georgia

ISBN 1–57312–034–0

Silver Linings
Triumph of the Challenger 7

June Scobee Rogers

Copyright © 1996
June Scobee Rogers

Peake Road
6316 Peake Road
Macon, Georgia 31210-3960
1-800-747-3016

Peake Road publications is an imprint of
Smyth & Helwys Publishing, Inc.

All rights reserved.
Printed in the United States of America.

The paper used in this publication meets the minimum
requirements of American Standard for Information Sciences –
Permanence of Paper for Printed Library Materials.
ANSI Z39.48-1984

Library of Congress Cataloging–in–Publication Data

Rodgers, June Scobee.
  Silver Linings: Triumph of the Challenger 7 /
  June Scobee Rodgers
      x + 114 pp.   5.5" x 9" (13.97 x 22.86 cm.)
      ISBN 1-57312-034-0 (alk. paper)
      1. Astronautics—Study and teaching (Elementary)–United
  States.
      2. Rodgers, June Scobee—Contributions in education.
      3. Challenger (Spacecraft)—Accidents.
      4. Loss (Psychology)—Religious aspects—Christianity.
      I. Title
  TL845.R63   1995
  363.12'465—dc20   95-36078
                                                      CIP

Contents

The Lessons Learned

The Shared Mission

High Costs

The Mission Continues

Foreword

Silver Linings is the true and inspiring story of the triumph of faith in the face of overwhelming defeat. It speaks an astounding message of undying vision and determination and love. What you will find here is not some Pollyannaish flight from reality, but instead an account of near unbearable tragedy that will make you cry and almost incredible triumph that will make you cheer. You will learn how the unseen power of God can change night to morning and breathe new life filled with hope, joy, and purpose into a desperate, sorrow-filled, and meaningless situation. You will see that faith can turn the darkest gloom into a bright cloud radiating with new life.

A fiery blast high in the cold, blue Florida sky on that unforgettable morning of January 28, 1986, tragically ended the flight of the *Challenger* space shuttle and claimed the lives of its seven courageous crew members. The explosion inflicted numbing grief upon the surviving family members and sent ripples of pain and disbelief throughout the world. As a moment frozen in time, the *Challenger* disaster engraved itself on the minds of all Americans, especially the children who were then of elementary school age awaiting their "lessons from space." The "education mission" of the *Challenger* seemingly taught us all the bitter lessons of disappointment, despair, and unfulfilled dreams.

Buoyed by their undying commitment to the vision of their lost loved ones, however, the wives, husbands, children, parents, brothers, and sisters of the *Challenger* Seven gave themselves to the task of turning tribulation into triumph. Driven by a determination to see the *Challenger* mission continue and sustained by the compassion and strength of family, friends, and God, the survivors of the *Challenger* Seven bound together to turn their "scars into stars." From the terrible dust of defeat has risen again the proud wings of victory. The *Challenger* mission continues, and the bright dreams of its crew live on in the inspiring work of the Challenger Centers for Space Science Education.

At the center of this heroic effort to transform tragedy into victory has been and still is June Scobee Rodgers, whose husband Dick Scobee was the *Challenger* commander. Dr. Rodgers is the founding chairman of the Challenger Center for Space Science Education and the "sparkplug" who has energized the efforts of everyone involved in the Challenger Center program. Anyone who meets June is immediately lifted by her excitement and conviction for the great work she is about, and few leave a meeting with her without finding that they have come to see their own tasks and lives in a newer, brighter light. Few also know the full depth of her great sorrow and the tremendous struggle she has undergone to discover new life in the midst of so great a disaster.

June's life has been characterized by a winning spirit that refuses to give up hope in the face of apparent defeat. Few of her elementary school classmates believed in her childhood dream to become a teacher. They knew she was from one of the poorest of the poor families in Alabama, where her itinerant carpenter father often moved the family either to rough, rural areas or inner-city government-subsidized housing. Their taunts that she would never see the inside of a college hurt her young heart, but they could not deter her from pursuing her goal. She eventually earned her doctorate and went on to conduct research at the university level. She has even served as a member of the President's National Advisory Council on Education and Improvement.

June's dream of teaching eventually merged with Dick Scobee's dream to fly. At first discouraged from pursuing his dream, with June's constant support Dick found a way not only to realize his dream of flying, but eventually to become part of one of the most ambitious and epochal flying endeavors in human history as the commander of a space shuttle. Everything seemed to fit perfectly for Dick and June when he was assigned to "the education mission" of the *Challenger* shuttle. June encouraged Christa McAuliffe in preparing the lessons that Christa would teach from space to thousands of children eagerly gathered in hundreds of classrooms. Dick worked hard with Christa and the other crew members to prepare for the rigorous flight. The *Challenger* mission offered both June and Dick the grandest opportunity they had known to fulfill their lofty dreams together.

The *Challenger* disaster seemed to have exploded all those beautiful dreams into a million pieces. But June's faith in God's unwavering love, and her staunch faith in the dreams she and Dick had shared with the other *Challenger* astronauts and families, would not let her give in to the waves of defeat that swept over her following the disaster. Miraculously, the spirit of Christ rekindled the embers of her dreams and gave them new direction. The Challenger Center for Space Science Education was born, and through the Center the *Challenger's* great mission continues.

From lowly beginnings and through wrenching struggle, June Scobee Rodgers has proven herself to be one of God's great winners. Her story has touched my heart and renewed my conviction that our God can turn all our scars into stars. May her story here in picture and word lead you beyond all your tribulations into triumph.

Dr. Robert Schuller

Preface

We have great cause to celebrate in America. Our nation is built upon a vital pioneer spirit that did not end with those early sailors who navigated across the Atlantic or frontier explorers who forged their way across our nation. Neither did it end with the fiery explosion of the *Challenger* space shuttle and the loss of its seven crew members. That pioneering spirit of the American patriot lives today because the frontier is still there, and because the will is in the hearts of great Americans and courageous people around the world who still believe we have a frontier to cross and a mission to continue.

This book is dedicated to the trailblazers who have gone before us to pave the way, and to the brave-hearted who accept the challenge to follow. It is dedicated to great teachers everywhere, who do more than impart knowledge of the past; they inspire dreams of the future. It is dedicated to the memory of the *Challenger* Seven astronauts (my husband and friends) who gave their all, and to our families, my dearest friends, and our children and their children who have all taught me lessons about life. Sorrow is balanced with joy, and struggle with reward. And like the sea, our lives can be turbulent or calm, but on that sea is a mighty power who loves and cares for us, and most of all believes in us.

The writing of this book was a journey in itself–not a grand journey, but more like a trail into the wilderness. At the beginning of the trail, my daughter, Kathie Scobee Fulgham met with me to help with plans, direction, and encouragement. Robert Schuller, an evangelist and the author of *The Be (Happy) Attitudes*, inspired the idea and encouraged the dream. Several fellow travelers stopped along my path to help or encourage the work.

Some people encouraged me to write not the generic story, but my story. Clay and Barbara Morgan encouraged me to write a book that could help others, to tell about my journey from tribulation to triumph. As friends, Vance Ablott, Danny LaBry, and Colleen Phillips helped me to locate information and provided their personal support. George Abbey and the team at NASA's Johnson Space Center helped me to locate photographs to support telling the story. For Challenger Center's tenth anniversary, Lee Greenwood and Dan Bradley wrote the story in song, "Written in the Sky." Famous artist Bob McCall helped to create the story in beautiful color on canvas and gave me permission to use many of his other paintings.

I met others along the trail who influenced me or assisted in the work. Some of them were Chip Bishop, Ellsworth McKee, Eric Rodgers, John Lowery, Helen Exum, Donna Boetig, Mike Carroll, Fran Simmons, Rick Hutto, Janice Ogrodowczyk, Nell Mohney, Mom and Dad Scobee, Dick Methia, Barbara Bush, Bill Lane, Grace Corrigan, Cooky Oberg, Bill Nash, E. Paul Torrance, and Louise McCall. When the trail of memories became too difficult to travel, one or several of my seven grandchildren's visits would brighten the day and point me back to my path and the trail that led into the future.

The staff at Smyth & Helwys helped me clear the path. Their magic and dedication turned the manuscript into a book that we hope can provide a little light to help fellow travelers in times of darkness find their silver linings beyond the clouds.

Near the end of the trail, my daughter Kathie and son Rich read the manuscript and encouraged its publication. Their tender words of support have made all the difference. At the end of the journey, I found this book and a loving husband, Don Rodgers, waiting for its completion. I am grateful not only for Don's patience and assistance in seeing the work completed, but even more so for his understanding and compassion on our own journey that together we travel through life.

Photo: NASA

The Shared Mission

*H*igh Flight

Oh! I have slipped the surly bonds of Earth
    And danced the skies on laughter-silvered wings;
Sunward I've climbed, and joined the tumbling mirth
    Of sun-split clouds—and done a hundred things
You have not dreamed of—wheeled and soared and swung
    High in the sunlit silence. Hov'ring there,
I've chased the shouting wind along, and flung
    My eager craft through footless halls of air.
Up, up the long, delirious, burning blue
    I've topped the windswept heights with easy grace
Where never lark, or even eagle flew
    And, while with silent, lifting mind I've trod
The high untrespassed sanctity of space,
    Put out my hand, and touched the face of God.

—John Gillespie Magee, Jr.

"High Flight" by John Gillespie Magee, Jr., appeared in the *New York Herald Tribune*, February 8, 1942, copyright IHT Corporation.

Magee was an American who was born in Shanghai, China, and who was killed at the age of nineteen while flying with the Royal Canadian Air Force in December, 1941, over England.

January 28, 1986
Kennedy Space Center
Florida, USA, Planet Earth

In silence, we watched from our viewing site at Kennedy Space Center the liftoff of our loved ones in the space shuttle *Challenger*. In only a few more minutes, I thought, they would be orbiting the earth, at work, confidently conducting the duties of their mission. For me, those anxious moments waiting for their launch into space were filled with fond memories and with questions left unanswered and problems left unsolved.

Photo: NASA

Friendship

Only moments before, the immediate family members of the crew had been gathered downstairs in the empty offices of the NASA officials. We were waiting anxiously for the astronauts (our husbands, wife, fathers, mother, sons and daughters) to climb aboard the orbiter and begin the countdown of the twenty-fifth space shuttle launch of Flight 51-L, the "Teacher in Space" Mission.

We knew how to wait. This was our third, early, crack-of-dawn morning trek out to the Cape eager to view the space shuttle launch of our loved ones and friends aboard the *Challenger*. We traveled by bus from our rented hotel rooms and apartments. Many of us knew each other well.

Lorna Onizuka and I had lived only a few blocks from one another in Houston, actually Clear Lake City, near NASA. For years before that, we lived just around the corner from each other

Photo: NASA

Space Shuttle Orbiter Crew Members for 51-L
(back row, left to right) El Onizuka, S. Christa McAuliffe, Greg Jarvis, Judy Resnik
(front row, left to right) Mike Smith, Dick Scobee, Ron McNair

Photo: USA Today

Seated (left to right) Jane Smith, June Scobee, and Lorna Onizuka
Standing (left to right) Steve McAuliffe, Cheryl McNair, Marcia Jarvis, and Chuck Resnik

at Edwards Air Force Base in the California desert. At first, I only knew her and her loved ones as the Hawaiian family whom I sometimes saw biking on the street or shopping at the grocery store. We were drawn closer when NASA announced in early 1978 that our husbands had been chosen as finalists in the selection of the first thirty-five astronaut candidates chosen to fly the space shuttle.

Ellison Onizuka was a test engineer. My husband Dick was a test pilot. They both took their work seriously and cared about their families immensely, and they both had similar interests in the space program. Dick admired El's fun-loving spirit and outgoing personality. He admitted to me once that he wished he could be more outgoing, like El. Both he and Lorna were fun people. They had a knack for turning a tense situation into a humorous one—often at their own expense. Lorna's friendship, her ability to make us laugh at ourselves during otherwise trying moments, endeared her to me and still does. We were blessed to have her with us. I needed her. We all needed her sense of humor to see us through those anxious early mornings.

Jane Smith was the other military wife and also the mother of three. She was a petite, blond beauty, a vivacious lady who had a lovely family. Her husband Mike was a naval test pilot who was selected after Dick and El into the group of astronauts. Dick admired Mike's experience and ability as a super pilot. Whenever Dick talked with me about his day of training in the flight simulator, he spoke of Mike with great respect and admiration. Together they felt the tremendous responsibility of commanding, flying, and safely landing the bird (their affectionate reference to the orbiter).

Jane was my kindred spirit. Both of us grew up in the South and married pilots, so we understood from a wife's point of view our husband's dreams of flying. We knew their passion for the military mission whether it was flying a new plane in a test program or training during practice missions, or the real battle of the Vietnam War. We lost friends who never returned from those missions. We stood alongside their wives and children at funerals and

Photo: NASA

T-38 Flyover "Missing Man Formation"

during moving ceremonial "flyovers," when a single plane would fly out and away from a formation of planes, representing the missing pilot, a lost friend.

More than a pretty lady, Jane was also compassionate and thoughtful; she was my soul mate that morning. We exchanged more all-knowing glances than words. Our common experiences and vicarious love for flying bonded us in a strong friendship. Jane, Lorna, and all the family members were wonderful company on this journey we traveled together. Our paths had crossed at this time; for that I was grateful.

Perhaps no other family member was more aware than Jane of the potential of the six-and-a-half million pounds of thrust waiting to be unleashed on the *Challenger* launch pad, unless it was our son Rich, a young twenty-year-old cadet in his senior year at the Air Force Academy. He stood there with his mother and sister while a few of his classmates and close friends watched from the viewing stands. Already a pilot and soon-to-be-officer in the Air Force, he knew more about the mechanics and aeronautics of flight than I did, even after all my years of flying "co-pilot" with Dick in our small Starduster II Bi-Wing airplane, or after twenty-six years of life beside a man whose dream since childhood had been to fly.

Rich Scobee beside Starduster II

Dreams

Dick's dream to fly, to touch the sky with "laughter-silvered wings," began as early as the time he could talk and ask his mother for a wind-up toy airplane he saw in the Sears and Roebuck catalog. By age three, he was ready to fly solo. At Christmas, his Aunt Tene gave him a toy riding plane with pedals much like a tricycle. It was his favorite toy, his parents told me, and it brought him hours of joy and merriment until he wore out the wheels from use. Then his dad made the plane into a swing and hung it from a tall cherry tree in their backyard. Soaring higher, he was a little closer to the sky where his child-sized hand could reach toward his star, but this small-town country boy was still far from ever flying the real thing.

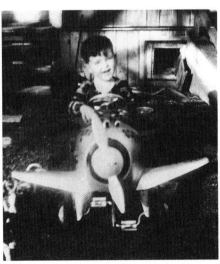

Dick Scobee "flying" toy airplane

During his teenage years, Dick's sights were still turned toward the heavens, to the stars. The love of airplanes and dreams of flying were his strongest motivation throughout his school years. Whether it was sketching airplanes on his notebook paper during class time, or building plastic and wooden models after school to hang on string from his bedroom ceiling, it was his favorite pastime. When he took me (his young bride) to his home to introduce me to his parents, he wanted me to see first his collection of airplane models, even before any prized high school trophies.

Dick's dream to fly airplanes went unfulfilled for years. He met obstacles at every turn. When he graduated from high school, he wanted to attend a military academy, but that dream was beyond his reach. A high school counselor told him he didn't have what it took to apply for an academy, meaning that he didn't know a senator personally who could recommend him. He joined the Air Force as an airplane mechanic, but his dream to fly went undaunted.

His persistence paid off. His dream was finally realized when, after graduating from night school at San Antonio Community College and receiving a degree in aerospace engineering from the University of Arizona, he was accepted to pilot

Photo: NASA

Dick Scobee with NASA F-111

training at Moody Air Force Base, Georgia, in 1965. In the next twenty years, he flew over forty-five types of aircraft and logged more than 6,500 hours of flying time as an Air Force officer, a combat pilot, a test pilot, and a NASA astronaut pilot.

Dick's dream to fly extended even into space. He flew the small experimental X-24B aircraft and the giant 747 shuttle carrier that ferried the spacecraft across the country. As pilot of the *Challenger* in 1984, he guided the spacecraft in orbit so the crew members could repair a broken Solar Max satellite. A mechanic at heart, he encouraged his fellow crew members to show up at an in-flight press conference in T-shirts that read "ACE SATELLITE REPAIR CO."

When he mailed his application to NASA to apply for the position of astronaut pilot, he wrote the following paragraph in response to a question about why he wanted to be selected:

> Why do I want to become an astronaut? Probably, my most compelling reasons for wanting to become an astronaut are a desire to extend and use the engineering and test pilot experience I've gained, to hopefully aid in the success of the space program, and for my own satisfaction in realizing a very longstanding personal ambition. I thoroughly enjoy being a test pilot and performing flight related tasks, and the astronaut program is, to me, a logical extension of that function into new frontiers. It is my belief that by the manned exploration and exploitation of the potentials of space and the planets, we satisfy a basic need of mankind to explore and probe the unknown, and I simply want to be an integral part of that exploration.

Years earlier, as teenagers on a date (we didn't have much spending money), we parked at the end of airport runways to

Art: Bob McCall

McCall art of Research Flight Center at Edwards AFB, California
Dick Scobee flew several of these planes.

watch planes take off and land. Dick taught me about aerodynamics, all about wings and airfoils and how they provided lift. We studied the stars and talked about astronomy, and together we dreamed of the day when he would fly and I would teach school. All we had were our dreams… poor country kids with big dreams.

Later when those dreams became a reality and I was teaching high school and college, Dick traveled on field trips with me or helped me to create a simulation to aid my students' understanding of science. Once we created homemade rockets to teach a physics principle.

That cold January morning in 1986, we watched and waited for the same scientific principles to be carried out, but this time it was for real with people I knew and loved, sitting atop a real rocket-powered space shuttle. The ultimate dream had come true: Dick Scobee was the commander of flight 51-L, the educational mission.

Faith

Cheryl McNair, a young loving mother, waited with us too. With help from her father, she stayed with her two young children to see her husband Ron, a talented physicist and scientist, fly on his second mission into space.

Ron and Cheryl were an admirable couple. They lived only a few blocks from us in Clear Lake City, Texas. Often we saw one or the other on the schoolyard track near our homes. We jogged miles of laps around that track most evenings and weekend mornings for years. I remember visiting Cheryl, a new mother just home from the hospital with her newborn son. She beamed with the newfound love of motherhood.

I admired Cheryl for her ability to juggle first a teaching career and then a job at NASA with managing a home and family. Most of all, I treasured her for her strong Christian faith, her beauty within and without. Cheryl taught me how to pray, how to really pray for the strength to see God's will be done. Though she was young, she was wise in the spirit of Christian love. She shared her faith freely this launch morning, not so much in words as in actions, graceful movements, and calm, melodic words of concern for the comfort of her children and their grandfather.

Judy Resnik's father was there with us. Judy was the unmarried *Challenger* astronaut, so her dad and his wife represented the immediate family. It was not Judy's first space flight, so I thought her parents were probably as much aware as I was of the forthcoming events. I knew of Judy's love for her father (a kind and professional sort of fellow), as did most of the nation.

On her first space mission, this petite beauty with dark, curly, flowing hair and sparkling eyes as dark as midnight painted a poster-board sign that read, "Hi, Dad." The space cameras beamed the sign back to earth and into homes of people all around the world. She had it all, I thought. She was a brilliant engineer-scientist Ph.D., an astronaut mission specialist, had a father who loved her dearly, and all that striking beauty. I enjoyed knowing Judy. Sometimes, Judy and I would meet for dinner at a favorite seafood restaurant in the harbor. We laughed about stories of our youth, but laughed more at each other trying to tell jokes.

Before that first flight, Judy was assigned to assist a newscaster on the evening news to explain the goals of an earlier flight. When I complimented her several days later for her special appearance, she quizzed me, not on how she looked or came across on the camera, but on how well she explained the mission objectives. Whatever responsibility she was called upon to undertake, she took seriously.

Missions

Judy Resnik was mission-oriented, not unlike the rest of the crew members. Like most astronauts, they were often quoted as saying they were ordinary people, doing a not-so-ordinary job, working together as a team on a mission, representing the pioneering spirit of the people of our nation.

During Dick Scobee's first space flight on 41-C, the "Solar Max Repair" mission, President Reagan spoke to the astronauts, commending them for a job well done. When "Scobs," as his friends called him, returned home, visitors and reporters were eager to meet him. Instead, he asked me to slip away with him to go to our favorite hideaway restaurant.

At the restaurant, we were seated at a table near a large glass window that allowed us to look out onto a clear lake that reflected the bright evening stars—stars that sparkled no greater than those reflected that night in Dick's eyes. He told me all about

Photo: NASA

Space Shuttle Orbiter Crew Members for 41-C
(left to right) Bob Crippen, T. J. Hart, Ox Van Hoften, "Pinky" Nelson, Dick Scobee

the excitement of flying weightless and the surprise he felt when, on looking back at Planet Earth, he suddenly realized how fragile and vulnerable it appeared. After dinner, I asked him if it didn't make him mad that the president forgot to mention his name along with the other crew members, while he was flying in space. He answered me with, "June, that was nothing. What was important was we got the job done; we completed the mission successfully."

Never had I seen such schoolboy wonder in his eyes, or smiles so broad, or stories told with such delight as those he relayed to me that night. Our conversation leaped from his experiences aboard the spacecraft to the great frontiers of both the inner mind and outer space to the need for missions in life. We wondered about feelings of responsibility and about whether or not we are destined or called upon to undertake certain missions or goals in life. That evening I concluded that life is probably so complex because God gives us "mission" tasks, and just when we think we've managed to complete one, He challenges us to another. We agreed that we need goals or missions to have direction in our lives, but the thrill, even greater than reaching the end, lies in learning to enjoy the journey along the way, moment by moment.

Photo: NASA

Responsibility

On the cold morning of the *Challenger* launch, we waited for yet another mission. I felt the greatest responsibility for Marcia Jarvis and Steve McAuliffe, the remaining immediate family members, whom I knew the least. Just as Dick worked to inform, protect, and promote the team spirit among his crew, I too felt a responsibility to their families, partly because that's my nature. The teacher in me was always concerned, caring, overly protective, and motherly—in sometimes annoying ways, I'm sure.

The role, though not a duty, of the commander's wife is usually carried over from one commander's wife to another. She provides receptions and get-togethers to talk about and plan for upcoming events in preparation for a space flight. It was my nature to go beyond, even overboard, on the nurturing, protective side of things to help everyone understand and feel comfortable with the anxious days prior to a flight and the days of celebration following a safe return.

I had lessons yet to learn about accepting responsibility for my own actions, much less those of everyone else and the outcome of every situation. I needed to learn that we can't control others or certain events in our lives, but we can control how we react to those events.

Without knowing it, Marcia Jarvis influenced me a great deal by how well she adjusted to new friends and new situations when her husband's crew assignment shifted. She had grown to know and feel a part of another crew family whose mission flew before ours. Her husband, Greg Jarvis, the payload specialist assigned from Hughes Aircraft, had been bumped off the

Columbia flight and onto Dick's flight because Congressman Bill Nelson from Florida wanted to take a ride into space. To this day, I don't understand that action or know who was responsible for the decision, but we all know the consequences and loss for Marcia.

At first, Dick was distressed about Greg's shift to his flight because he knew the craft was exceeding weight limits due to the experiments Greg would bring on board. Dick felt an overwhelming responsibility for the decision, he told me, and was angry that he wasn't consulted as the commander. Although he complained and balked to his supervisors, to my knowledge, the only change that took place was a waiver for the added weight.

We all loved Greg. What a great guy! What was there not to like? He was serious and studious, yet a fun-loving friend and caring husband. Dick liked Greg and told me often how much he hoped they would remain friends after Greg returned to California with Marcia. That morning Marcia was one of us—waiting, interacting, smiling, and laughing as only Marcia can with her charm and self-effacing nature.

I felt the most concern for Steve McAuliffe and his children. I knew this bright attorney and former military man could manage for himself, but perhaps I sensed he was least prepared for what was about to take place. Also, in the year of crew-training, I had grown to know Steve and their children through loving stories told to me by his wife Christa when she visited me in our home or when we met for lunch or dinner.

Though we were both classroom teachers, I knew Christa first as a mother. We worked and planned together to create Halloween costumes "long distance" in Houston for her to take home to Concord, New Hampshire. After her family flew out to see her and visited Johnson Space Center, she proudly shared anecdotes about her children's first impressions.

Holidays

As a mother, I knew and understood how much Christa missed her family. So during after-dinner talks with Dick, I would slip into the conversation tidbits of information about upcoming holidays and how much Christa's children missed her. Although Dick was the most mission-oriented person I knew, he had a soft spot in his heart for family. He would respond to my hints with a wink, then tousle my hair and volunteer a remark something like, "O.K. I'll tell Christa to make plans to fly home to her family." Even though the crew often worked through evenings and weekends, everyone usually had their family with them in Houston except for Christa.

Christmas was a special holiday that year for all of us. We were excited for the crew, our loved ones, and our friend Christa. Energy was already high, a fact that made the holiday season even more festive. Before Christa left for Concord, she and Barbara Morgan, the back-up teacher and finalist who would take Christa's place in case Christa couldn't fly, came to my house to have dinner with the spouses of both this *Challenger* flight and the spouses of the *Columbia* crew that had just flown.

Christa and Barbara wanted to help, so they offered to bring a Christmas wreath and decorated cookies. Those two were as creative as they were lovely, and they celebrated with us in great fashion. When we posed together for photographs that evening, a wife from the other crew said that we were all little short things, not much bigger than minutes. It was true, not one of us was much over five-feet-two or three inches except for Christa who stood tall compared to us at five-feet-six-inches.

Standing: (left to right) Jane Smith, Cheryl McNair,
June Scobee, Lorna Onizuka
Sitting: (left to right) Christa McAuliffe and Barbara Morgan

Christmas Eve was upon us before we knew it. I was bursting with joy. Our son Rich had come home from the Air Force Academy for the holidays, and our daughter Kathie brought more than gaily wrapped packages. She brought her ten-month-old baby, Justin, to us. Our first Christmas as grandparents—how we relished those days together!

Dick and June with grandbaby Justin

Excitement was mounting for other reasons too. We were expecting special guests for Christmas Day dinner. Kathie was a treasure to help in the kitchen when she wasn't managing the baby. Married now and a young mother, she was interested in creating her own holiday traditions, and the most important one for her that day was to learn how to make Grandma's southern cornbread stuffing for the turkey.

We began making preparations for the Christmas feast, chopping and stirring at the butcher block table in our country kitchen. As the smell from the oven drifted out to the others, it beckoned them away from what occupied their attention to join us and help with basting the turkey, mashing the potatoes, and setting the table with the good china and crystal that Dick had brought home from his travels to Germany. In our off-key voices, some of us sang Christmas carols. All of us told stories about our lives apart, mixing in favorite memories of our lives together.

Time slipped up on us. Before we were ready, the doorbell rang, and the guests arrived. Like Santa himself, Clay and Barbara Morgan greeted us at the door with presents and good wishes. We enjoyed most their friendship, a priceless gift that Christmas Day. Special guests and festive meals always encourage marvelous, cheery, nostalgic conversations, but this year the conversation turned just as often to the future and expectations of adventure and promise.

Clay, a novelist and forest service smoke jumper, had been able to join Barbara in Houston as she trained to be Christa's backup, escorting her to T-38 practice flights and evening events. Dick and I had grown to really love Clay and Barbara and felt as though we had always known them. A few evenings, our philosophical discussions about the workings of the mind within and our solar system beyond took us into the wee hours of the morning. Because our lives were so busy with Dick's training and my teaching, we promised ourselves and each other that this friendship would continue, and that someday soon we would visit McCall, Idaho, where

Photo: NASA

Barbara Morgan, Mike Smith, Clay Morgan, Christa McAuliffe,
and Dick Scobee walking from hangar at Ellington AFB,
Houston to T-38's for flight training

they lived together in a log cabin on a mountain lake—quite a contrast to the space world we lived in, I supposed.

Dick worked through most of the holidays preparing for his flight, but Christmas Day was set aside for family, and New Year's Eve for me. No great celebration; instead, Dick planned a little time for me to interview him for an article I was writing to be published in the March, 1986, issue of the National Science Teachers magazine, *Science and Children.*

Someday, we planned, we'd write a children's book together about cats and dogs in space. We laughed as we imagined a cat with legs outstretched and tail fluffed in fear anticipating the end of the free fall and the dog floating in the weightless environment with three times as much floor to sniff. We mused about the antics and shenanigans of these two unnatural choices for *Pets in Space.* For the article, however, he suggested to the young readers that he would build a special carpet-covered pole across the spacecraft for the cat, and that both animals would have to wear some form of diaper. The laughter and craziness from our play brought balance and levity to our otherwise more serious days of training and preparations.

Pioneer Women

When Christa returned to Houston after the holidays, we found time to visit on several occasions. We were both great talkers, teacher to teacher. We shared classroom stories about our students, our concern for them as people and our keen interest to see them reach their greatest potential. We believed in hands-on, real-world opportunities for our students.

Recently, she told me, she had taken her class on a field trip to learn about the judicial system in a court of law. She had also created a curriculum about pioneer women, juxtaposing the westward movement with pioneer women of today flying into

Photo: NASA

First Women Astronauts
Counterclockwise: Rhea Seddon, Kathy Sullivan, Sally Ride,
Judy Resnik, Anna Fisher, Shannon Lucid

space. Christa was a pioneer in her own right. She wanted to power her own Conestoga wagon with rockets, keep a diary or journal about her voyage, and tell the story to all of her students—those back home in Concord and those around the world waiting for lessons from this beloved schoolteacher.

Photo: NASA

Christa in KC-135 trainer called "Vomit Comet"
because of 0-gravity, weightless exercises
Simulated flight of Christa preparing to teach lessons from space

Teachers

One evening I invited the students from the college classes I was teaching to my home to review and comment on a final project. Christa came with Barbara. They wanted to meet my students and congratulate them for their efforts in creating a delightful children's book. The book is about a curious little tinkerbell-type bug who slips aboard a shuttle flight as a stowaway, then tells her story of adventure and conflict to young readers. My students titled their book *The ShuttleBug*. It was a wonderful evening and a great opportunity for my students (all teachers themselves) to share their project with the "Teachers in Space," Barbara and Christa. In turn, we all enjoyed hearing about their exploits, training, and hopes for the upcoming space flight.

As a remembrance of the evening, I gave to each person there a small, framed piece of art, matted in pastel colors with a

Students presenting *Shuttlebug* book to Barbara and Christa.
Barbara and Christa are seated on floor.

phrase written in calligraphy—one with which we as teachers all identify—

I touch the future...I teach.

—Anonymous

Those teachers and I, and teachers from around the world, would fly in spirit with Christa because she represented for all of us that a classroom teacher could have the "right stuff."

More than just to educate from a spacecraft, we thought Christa could affect the course of lives for teachers. We hoped that people around the globe would recognize and honor the educators who had opened the way for their students to discover talents and awakened in them the desire to learn and the courage to dream. Christa wanted to explore space, inspire children, and teach lessons about the space frontier. The dreams of thousands were riding on Christa's shoulders.

Like any dedicated teacher, Christa had lessons to plan and preparations to make for her orbiting classroom among the stars. Not unlike environments in space, classrooms can have both calm and turbulent days, so Christa was prepared not only for the joys and rewards of teaching, but also for confrontations at the workplace. Whenever she told me about frustrations, I tried to help her get around the obstacles—though she could have managed quite well without me. Once I encouraged Dick to talk to some of the folks who were causing the difficulties, but on another occasion, I couldn't help much, although I sympathized with Christa when Dick got upset about an emergency egress/escape exercise.

Before we knew it, the day arrived for us to pack our bags for Florida and catch the NASA passenger plane to the Cape. We were all there, except for the astronauts who flew ahead of us in

Photo: NASA

Five crew members of Space Shuttle Mission 51-L
receive slidewire escape training.
(left to right) Ronald E. McNair, Gregory B. Jarvis, Christa McAuliffe
(back left to right) Judith A. Resnik, Ellison Onizuka

their NASA T-38 training jets. Christa and Greg were wearing their astronaut flight suits. Lorna, Jane, Cheryl, Marcia, and I were dressed in casual clothes. We waited together for everyone to arrive before ascending the ramp to the plane. The sun was shining so brightly that January morning that we had to squint to see across to the flight hangar. I had an armful of roses to present to each of the ladies as they arrived.

We made anxious gestures to each other and the crowd of visitors and reporters who waited to see us off and waved. Several of the reporters had become my good friends, and one of them,

Wives of *Challenger* Astronauts at Ellington
(left to right) Cheryl McNair, Jane Smith, Marcia Jarvis, June Scobee, Lorna Onizuka

John Getter, was joining us later at the Cape. "Hey, June," he called, "Bet you wish you were the teacher taking the shuttle flight." "Sure do," I remember calling back to him, "and I'd take you with me too." We laughed, for we both had admitted to the other that we would take a space flight in a heartbeat, especially if NASA had a need for a grandmother. We also knew that Christa was the perfect teacher for the mission. She was animated, enthusiastic, intelligent, an adorable young mother, and a special friend to all teachers.

Christa "The W-R-I-T-E Stuff"– sitting across from June on NASA plane

Sense of Humor

The flight to the Cape was uneventful except for the fun we had. We were very much ourselves, friends on a journey traveling together. We talked seriously about logistics of the rendezvous at the Cape and other events with our loved ones. Not so seriously, we compared shoes and socks. Christa was wearing tennis shoes rather than the big, black brogan boots that Dick usually wore. That was smart, I thought.

We all gathered in the center of the plane to talk and tease one another. Most of us sat around a table, facing each other. Christa sat across from me and answered letters to her students and friends. She was full of promise, filled with great expectations and hope for her flight. I respected her for many reasons, but at that moment I admired her for her spunk and thoughtfulness. While she wrote her notes, I watched and then said playfully, "Christa is the *write* stuff!" They all passed their roses to me in a mocking congratulatory manner to tease me about the pun. What fun! They were all good to put up with me.

June gladly receiving roses and enjoying laughter

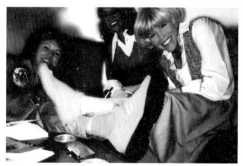

Christa showing her "boots" to Cheryl and Jane.

We arrived at the Cape without a glitch. Dick and his crew were there too. I ran to greet him with a hug and kiss. Then, I heard a call to Christa from the crowd of people behind the gate. It was Steve McAuliffe. I couldn't wait to meet him. As I turned to welcome Steve into the group, Dick reminded me that Steve needed to pass his Primary Contact Physical prior to joining the crew. Steve knew that too, so he waved hello, and standing away from the crew and spouses, he tossed a T-shirt with the New Hampshire state seal on it to Christa, who passed it on to Dick for a photograph. It was as though I had always known Steve. In the next days we would have many opportunities to get better acquainted. We would be good friends, I predicted. I liked his great sense of humor—his light-hearted teasing and fun-loving expressions.

At noon the next day, the adult family members who had passed their physicals joined the crew members for a picnic at a cabin on the beach, near the shuttle launch pad.

Photo opportunity for *Challenger* Seven after arrival at Kennedy Space Center prior to their shuttle launch

The families gathered around the picnic tables on the deck of a house that hung out over sand dunes. Many met each other's extended families for the first time. Christa's parents met the Scobees. They shared in the excitement of the day and the anxious moments waiting for the shuttle launch.

After visiting with their families and eating fried chicken and potato salad, the seven crew members said their goodbyes to all but their wives and husband, who were asked to stay longer.

Dick called Kathie, who had a slight fever and could not attend the picnic, on the telephone to say he was sorry he hadn't been able to see her. She was feeling a bit gloomy, but sent her hugs and kiss through the phone lines and said something funny to make her daddy chuckle. As Dick and I left the cabin to walk on the beach, we agreed that we admired our daughter's resilience and her uncanny ability for spontaneous humor. Traditionally, it's children who learn from parents. In our case, our children were growing up well in spite of us, and we were learning much from them.

Laughter is a great release for tension, I thought, and so are the natural sounds created by nature at the seaside that drew us to the beach for a walk in the sand. The weather was unseasonably cool, but the walk gave us the chance to be alone. Dick pulled me close to him as we stepped out on the wet, sandy beach, leaving behind only a faint track of footprints at the ocean's edge.

The salty sea breeze, the sounds of hungry sea gulls, and the drifting bubbles at our feet were restful images. Most of all, we enjoyed the steady, rhythmic (even therapeutic) sounds of the ocean waves crashing upon the land, then gently flowing back and lazily churning under and into the next. . . .

We decided to walk up the coast until we could see the shuttle perched on the launch pad waiting for liftoff the next morning. When we had the first glimpse, we stopped and gazed longingly out across the sandy marshland to the shuttle cloaked in the misty air. Finally Dick spoke, "That's home away from home for the next week for the seven of us. . . . It's going to be crowded!"

A final walk along the beach together.

We smiled, then turned to the sea and stared out at the horizon, across the Atlantic Ocean. Dick hummed a familiar tune about a sailor on the sea saying farewell, then turned to me and sang, ". . . for you are beautiful, and I have loved you dearly, more dearly than the spoken word can tell. . . ." We embraced. I felt his love and appreciated his tender words. I hung on to them for a long, long while, then teased, "You and Roger Whittaker could sing a great duet." He grinned, and took my hand as we turned back to return to the others waiting at the cabin.

On the morning of the launch, the weather was freezing cold. We looked out at a clean blue sky that served as a magnificent backdrop to what appeared to be, from our distance, a small replica of the greater-sized shuttle. It glistened in the light, all white and sparkling, perhaps because of the bright Florida sun and the ice that hung from the platform, launch pad, and shuttle.

Photo: NASA

High Costs

Risks and Rewards

When Dick called me that morning for another farewell, I asked him if they could launch with all the icicles hanging everywhere. He told me the engineers had knocked off the icicles that might cause a problem. I asked about the engine too, although I knew the jets that Dick flight-tested gained their greatest altitude in freezing temperatures. "Yes," he added, "they showed us pictures of the rockets blasting off in snow … said it was safe." We said our good-byes again, kissed into the phone, and went our separate ways—he to leave with his crew from crew quarters and take the bus to the launch pad, and me to take our bus with our families and friends to the viewing sight.

Riding the bus in the early-morning, still-dark hours, thoughts bounced in my head like my body bounced on the hard seat in the back of the bus. Who were the "they" in whom Dick had placed his confidence? Were these people the same "they" with whom Dick was frustrated over the weight limits? Although I was anxious, I worked hard at appearing calm, allowing tears to fill my eyes only for a moment when my daughter looked at me with concern on her face.

Rich exits family bus at Kennedy Space Center.

Was the concern for her dad, or for me, or for her anxious moments of exhaustion from caring for her infant son suffering with an earache? The previous night we had taken him to an emergency clinic for a prescription for antibiotics to help relieve the infection. Poor dear, I thought. My little girl, all grown up, holding her own baby—her daddy's delight; "Pooh Bear," he called her. Was she prepared for this flight? Caring for a baby, concerned for her dad. Had I helped her, talked to her, prepared her for the unexpected?

What about all the other children? I asked myself. Counting our grandson, twelve children and teenagers were with us that morning on the bus. I felt uneasy and an overwhelming amount of responsibility for them all. Were they informed? Did they know? Had their parents told them, talked to them? Could I, should I say something else, something more? Already I had talked with Christa about the safety and the risks. More importantly, Dick had talked not only to her, but to all of the ten teacher finalists about the risks of space flight.

Less than a year ago, Dick had walked into the house after work, put his briefcase on the kitchen table, and asked me with his boyish grin, "What do you think about your husband being the commander of the 'Teacher in Space' flight?" He grinned because he knew my response. Ten teachers had been selected. One of them would fly aboard a shuttle, and Dick wanted to know what I thought.

What a perfect choice, I remembered thinking. After all, an astronaut married to a schoolteacher would have great compassion and understanding for another teacher. Dick Scobee was my best guest speaker for the Texas A&M University Summer Space Camp that I directed. He would be the perfect commander. Our career paths would cross with that mission. In the past, Dick had helped me. Now, I could help him with his additional, educational mission—to take a teacher into space—to inspire, learn, and touch the future.

Dick wasn't as excited and asked my opinion about his flying to Washington, D.C. to meet with the teacher finalists. He was to tell them that the chosen teacher must realize that space flight is still in the testing phase, that it is risky business and not like an everyday, routine airline flight. Of course, I encouraged him to say all those things and more, which he did. Later Christa and I talked about that day, as did other finalists who told me how important it was to them to meet the commander and hear his words of concern and advice.

The subject of safety issues came up again in early January when Christa was visiting our home with other guests, the Obergs, who had just published a book about space travel that included the topic of death in space. After our other guests left that night, Christa asked me how safe space travel was and if I would fly. We agreed that anything worthwhile in life was a risk. There had been twenty-four shuttle flights with astronauts returning home safely. "Yes, I would fly," I told her. Without risks, there's no new knowledge, no discovery, no bold adventure—all of which help the human spirit to soar.

I was an idealist, and Christa was my hero, not only because she was willing to fly aboard the shuttle, but even more so because she was aware of the risks. We agreed, though, that not everyone would embrace our philosophy. Then she confided in me that her opportunity to fly in space was important not only for

Photo: NASA

Christa and Barbara: Teachers in Space

her to teach her lessons, but to bring attention to all teachers who strive daily on their own missions to help children learn, reach out to others, advance civilization, and touch the future. She said she would talk to Steve again about her participation in the mission.

That morning, riding out on the chilly bus with all of those dear children, I wondered if Christa had found the time to have that conversation with Steve. Did they agree? Had they discussed the "what if's?" or the worst-case scenario? I shifted on the hard seat and questioned my reasoning. Second guesses? I had told Christa I would fly. If it was safe enough for my husband, it was safe enough for me. I prayed for us all—for our loved ones waiting to be launched, for those of us waiting to watch the launch, for our children, for children around the world waiting for lessons to be taught from the classroom in space.

Vehicle Assembly Building with Launch Control Center wing, the viewing site for families

Waiting

When we arrived at the Launch Control Center, the families departed the buses and were escorted into the offices that were traditionally set aside for the immediate families of the crew. The rooms had great wide windows that looked out to the launch pad, perfect for viewing the shuttle. The families waited, drank coffee or juice, and ate fresh fruit and breakfast rolls. We knew how to wait, how to chat about anything, everything, to help the anxious hours pass, especially when flights were delayed.

We looked out the window. We watched the TV news announcers, the NASA select channel, and our children. Some of them sketched at the blackboard, others stared out the window, and still others talked with the adults and helped to entertain the toddlers playing with toy cars on the desks or floor.

We all waited. The clock ticked away each moment as though it carried a heavy burden. Finally, the long-awaited final countdown began. We picked up the babies and cameras and climbed the stairs to the rooftop viewing area. We were accompanied by John Casper, the astronaut office liaison officer who had been with us throughout our trip. He was there to assist us in case of an emergency, but more than that he was a friend. Dick and I had known the Caspers for years also, since they had served together as test pilots at Edwards Air Force Base. It was comforting to have such a capable and dear friend with us.

Grouped together on top of the building, my children and I stood with Steve and his children. As we took photographs, I explained what would be taking place as the shuttle lifted off. We cheered as the solid rocket boosters ignited, and the shuttle carrying its precious cargo lifted off the pad. Only a few anxious moments were left.

We watched in silence as our loved ones climbed the sky sunward. Their craft from the distance seemed to sit atop a great flume of smoke. The floor shook with the sheer raw power of the millions of pounds of thrust. Dick had described the sensation of that part of the ride as a thunderous, runaway freight train. I imagined Dick in his ever-so-calm, matter-of-fact, take-charge mode. I imagined Christa in her excitement, nervously waiting for the solid-rocket boosters to separate, the engines to cut off, and the buoyant lift of weightlessness to signal their safe arrival into earth orbit.

Art: Bob McCall

The Unexpected

My son lovingly and protectively put his arms around me and his sister. As I reached to help my Kathie with the baby . . . it happened! The unspeakable happened. Standing there together, watching with all the world, we saw the shuttle rip apart. The SRBs went screaming off on their own separate paths, and the orbiter with our loved ones exploded in the cold blue sky, and like our hearts it shattered into a million pieces.

In stunned silence, we looked to each other . . . for answers, for information, for hope? What happened? No words came, no answers, only glances, only eyes spoke those next helpless minutes. I saw the pain in my children's eyes. If only I could turn back that clock, if I could stop time. I had no power, no answers. Steve McAuliffe's eyes caught mine. I was helpless. What could I say?

Oh, God! It can't be, I whispered to myself. I won't believe it. Why did you let this happen? My husband. No! No! Not my husband. Not his flight! Not his crew! My friends! Why, on my watch, did this atrocity happen? I let these people down. I encouraged the flight. I explained away the risks. Why God? Why them? Why us? Why me? This is not what happens at a shuttle launch. My son took my arm to support me. He helped me down the stairs. My legs felt numb. They wobbled clumsily. I stumbled. Finally, I spoke out loud, "What about the others?" I asked. "Who's helping the children?" Oh, no! Dear God, all those children!

Please God, keep me in one piece, for the sake of my own children, for the others. Make me steady; keep me calm. I patted arms, held hands, and whispered, "They'll be rescued. They'll be alright." I wasn't prepared for this role. No one ever is. Minutes later, we were on a bus to crew quarters. John Casper helped us. My memory fails here. It's all a blur like a nightmare that you try to piece together, but the fit isn't rational. I prayed for a miracle. Just let them survive, dear God. My head knew they were dead; my heart did not.

At a stop light, I turned my rigid body to the window to look out. Cars were everywhere, stopped on the streets, on the curbs and sidewalks. People embraced one another, perhaps friends, maybe strangers. Some stood outside their cars, heads resting on their arms across their doors. Others sat sobbing into their hands at the steering wheel. A wave of shock jolted across the land, through the people, and around the world.

Tragedy

The fifteen-minute ride to crew quarters seemed an eternity. No nightmare to wake from, no answers, no husband. Frozen with anguish, I pleaded, God, please don't let me cry. What do I say? How do I help these people? I must be strong for the others. My thoughts careened between the others and always back to myself. My life as it had been, the path Dick and I traveled together for twenty-six years, had reached its end, and the path that reached into the future with us at each other's side was gone, vanished. From that moment, I would be changed, a different person . . . alone. Dick's mission, his dreams, his life, his friends—those were all gone. They were the best of us, doing the best of things for their country, pioneers all, crossing the frontier into tomorrow.

The bus delivered us to the door of crew quarters, where we had said our good-byes the night before, and where we would say our last farewell to our loved ones. NASA officials, medical support and friends, and later our other relatives joined us at the quarters, an apartment-like complex. Gathered together in a central area, we were officially given the tragic news. "All crew members are dead. They could not have survived." No hope, no miracle, no chance. I left the others to slip away into Dick's private room so I could be alone to cry out in rage. Instead, I reached for his clothes still hanging in the closet and held them in my arms, fell to my knees, and sobbed incoherent utterances.

I wept for my husband, for myself and family, for our friends with us, and for those who worked for NASA and the contractors, pioneers themselves who supported and believed in space exploration. What a tragedy! What a terrible loss to humanity.

With that, I recalled studying and teaching the Greek tragedies and Shakespeare's plays. I had used the word "tragedy" so glibly for those lessons. I would parrot, "A tragedy is about a series of unhappy events that usually end in a disaster. In them, the heroes contend against a fate they cannot escape. It affects the people closest to the event but also sweeps across whole countries, affecting an entire nation of people." People—I was reminded about the others. I splashed cold water on my face, brushed back my hair, and put aside my own selfish thoughts to return to my family members to help them.

Message from Space

When I turned to leave the room, I saw my husband's briefcase and took it out to our family—my children and his parents. It was locked. I couldn't remember the combination, but my fingers did, and we unlocked and opened it. There inside were his personal belongings: a wallet, his keys, pictures of his family, shuttle souvenir pins, business cards, astronomy charts, flight manuals, an unsigned valentine card "For My Wife," and a scrap of paper with his handwritten note on it.

My children and I read the note as though it might be a message left for us. In a way it was. It was not the miracle I prayed for, but it was a message. In a way it was the answer to the question why: Why was it so important for him to fly into space? Why was he willing to risk his life? The note was a passage taken from Bob McCall's book *Vision of the Future*, written by Ben Bova, one of America's great space authors. I passed it to Dick's mother to read. She was holding Dick's light brown jacket, which he had given her only the day before to keep her warm. Cradling the jacket in her arms, she pulled it closer to her heart and read the note written in her son's own hand:

> We have whole planets to explore. We have new
> worlds to build. We have a solar system to roam in.
> And if only a tiny fraction of the human race reaches
> out toward space, the work they do there will totally
> change the lives of all the billions of humans who
> remain on earth, just as the strivings of a handful of
> colonists in the new world totally changed the lives of
> everyone in Europe, Asia, and Africa.

Dick Scobee
Commander, Space Shuttle Challenger

WE HAVE WHOLE PLANETS TO EXPLORE. WE HAVE NEW WORLDS TO BUILD. WE HAVE A SOLAR SYSTEM TO ROAM IN. AND IF ONLY A TINY FRACTION OF THE HUMAN RACE REACHES OUT TOWARD SPACE, THE WORK THEY DO THERE WILL TOTALLY CHANGE THE LIVES OF ALL THE BILLIONS OF HUMANS WHO REMAIN ON EARTH, JUST AS THE STRIVINGS OF A HANDFUL OF COLONISTS IN THE NEW WORLD TOTALLY CHANGED THE LIVES OF EVERYONE IN EUROPE, ASIA, & AFRICA.

Had Dick left the note in his briefcase for us to find if something happened? Did he write it on scratch paper to use to quote in a speech? All we'll ever know is that when we most needed a message, it was there. He left for us his dream for the world, his vision for space exploration.

Just then, we heard the announcement that Vice President Bush and two senators had flown in from Washington, D.C. to give the families their condolences. The vice president stood alongside Senator Jake Garn from Utah, who had been a pilot and flown a space mission, and Senator John Glenn from Ohio, the first astronaut to orbit the earth during the early Mercury missions.

Art: Bob McCall

Compassion

Three great men, holding back tears, somehow found words to speak for the president, for themselves, and for people across our nation to say they shared in our loss and great sorrow. Their kindness taught us about compassion. Silence! We stood together, the families circling the room, facing these men who suffered with us. If it had been only the day before, the commander would have greeted and thanked these national figures. No one responded. I looked to my children and then to Jane, who nodded to me as if to encourage a response. What could I say? I thanked them for their kind words, and still clutching the Ben Bova passage, I pleaded for them to keep space exploration alive—to continue the space program. Before they left us, the memory of a simple gesture stands out. Vice President Bush slipped us a small piece of paper with a home phone number and this message: "Call if you need us."

That night, NASA arranged for the families to return to our homes. Waiting to meet my family when our plane landed in

Sketch of Scobee home in Houston

Houston were astronaut Norm Thagard and his wife, Kirby. Norm had been Dick's close friend and shared an office with him at work. Kirby and I were also good friends. We had both taught together in the same high school. Others were there too. They held us in their arms, drove us to our homes, and took visitors into their homes. They helped. Where there was confusion, they brought order. They made arrangements for us; they protected us.

I walked; somehow I put one foot in front of the other. I talked and feigned strength, but inside I too was dying, stunned and uncomprehending. Worst of all, I thought myself gone—the nurturer, caretaker, provider couldn't even help my own suffering children. I mourned the loss of my husband who was my partner in life, best friend, and companion. I mourned the loss of myself. I was different, changed, a *widow*—I choked on the word.

Guilt

My feelings of guilt, both irrational and justified, kept crashing through my head like a thunderbolt. Why didn't I stop him/them from flying this morning? Why hadn't I prepared the children better? I grew weary of self-recrimination. Finally, I slept.

Upon waking, I opened my eyes to the familiar surroundings in my bedroom, to the light filtering through the lace curtains, the dusty rose and cream-patterned wallpaper, my husband's desk. Where was he? What . . . ? Oh, God. No! No! Like an electrical charge shooting through me again, I was awakened to the horrible, shocking reality of the day before, of the tragedy and great loss, not only for me and my family, but for the other families, our friends at NASA, and our nation and its children waiting for lessons. Helplessly, I cried. Purposefully, I prayed for strength and forgiveness.

Memorial services, arrangements for visitors—I made decisions mechanically during the following weeks. I began to acknowledge but not accept the deaths. The crew members had gone to a better place, or they were here with us, helping us. Their bodies were buried at sea, but they were here with us—angels among us, guiding us.

Nearly three months after their deaths, we learned that the bodies of the *Challenger* Seven had been recovered. Disbelief, pain, and now anger returned. Our loss was so public. Was there no sacred privacy? We had buried them, or at least I had rationalized to myself that they were at peace. The anguish returned, along with the media to my door and on the phone. Questions—some sensitive, some not—regrets, guilt, and sorrow returned like a giant, crashing wave, knocking me down. Drowning in my own self-pity and sorrow, my guardian angels Barbara and Fred Gregory (wonderful friends and neighbors and a fellow astronaut family) took me into their home and cared for me away from the public, the media. They fed me and put me to bed.

My husband was buried at sea, at peace, I cried to myself. They dug him up and all our sorrow as well for us to bury again. It's so public—this open, gaping wound is so public. I hated this life. I was angry with God for letting this happen. The energy used for anger, hate, and grief was sapping me of my life. I longed for privacy, even for death, to see Dick again, to see this painfully public life gone. I prayed for death in my irrational state. I threatened God to take me. If you won't take me, I begged, then give me strength to live this life, and help me to solve these problems and overcome these feelings of guilt.

Photo: Paul Hosefros/NYT Pictures

Coffin of Dick Scobee being carried from plane at Dover Air Force Base. Remains were returned to the family for burial.

New Life

This threat jolted me back to the living. That moment, I became like a child again. God was in control—not I. For the first time in my life since grappling with losses in my childhood, I relinquished complete control to God. A joyous spirit challenged me to live, accept my problems, and discover new joy awaiting me in a new life. I felt ashamed for my groveling, and I felt relief. For the first time, I really didn't feel alone. My faith was renewed.

The next morning, I was left alone, but I didn't feel alone. No longer the master of my own fate, the simple innocence of the child I once was told me that God was with me, in control. The pressure, the anger, the pain and guilt drained from my body like sap oozing from a tree. A part of me had died; a stronger, more centered and saner self was born. Into the mirror I stared, and through different eyes I saw a new person. My cheeks were flushed pink like the little girl I once was. The numbness I had known for months subsided. Piece by piece, I became more alive.

I stepped outside onto the shady, green lawn and thanked the early morning light for chasing away the night, the darkness. I looked at my hands, my arms. The tense, tight feeling was gone. I was more alert, more aware of my body, more conscious of life around me. I felt, for the first time in months, the tingle of a breeze floating across my skin. A ray of sunlight fell warm across my back. Sounds consoled me—birds singing from tree-top perches and children at play, laughing in the distance. A single golden daffodil bent forward as if to welcome me to spring, to new life.

A car drove up, and out ran my grandson. A year-old toddler fell pell-mell into my arms. What beauty this life had to hold!

What joy! How selfish, I thought, to want to leave this—to leave them. As I embraced my daughter, following close behind on the heels of my grandson, tears rained down my cheeks, but they were not tears of sorrow or self-pity—not this time. Lord knows I'd shed enough of those. No, these were tears of joy for life, for new life, for a rebirth to live out this life with my children and their children on a new journey, in a new direction, along whatever path God unfolded for me.

Funerals

Funerals, services, and burials were arranged during the month of May by seven families who laid their loved ones to their final resting place. Each family planned according to its religion or family tradition or the request of the loved one: Protestant, Catholic, Jewish, Buddhist; some private, others public.

I grappled with Dick's request to be buried privately and quietly in an old apple box and with his early thoughts about cremation. I talked to our children, Dick's parents, and our pastor. We chose to honor the man who relished his privacy. Already, a part of him was left buried at the crash site, his ashes scattered at sea on the Florida coast near the site of Cape Canaveral that he so loved. The part that was recovered from the wreckage, our children and I buried quietly and privately, along with his astronaut pilot wings, in the wooden box he requested. A modest headstone, symbolic of the plain and simple life he most wanted, was placed on his grave site near the Challenger Astronauts Monument at Arlington National Cemetery.

We also arranged a service for the more public figure—the astronaut, the military man who loved and served his country well. Friends and family joined us for this more formal ceremony. They walked with us from the chapel to the gravesite to pay their last respects. As a bugler sounded taps I placed flowers on Dick's grave located across from the grave of the Unknown Soldier, on a spread of grassy green beneath shade trees that reach to the sky just beyond the Washington National Airport flight pattern.

Photo: Washington Post

June places flowers on Dick's grave
at Arlington National Cemetery

Those were only the grave sites for Dick's body, laid to rest on May 19, 1986, forty-seven years after his birth. He's not there, though. He went with his six crewmates on January 28 when they climbed a golden beam of light sunward to the heavens to touch their stars, but instead put out their hands and "touched the face of God."

Adversity

Life moved forward, more certainly after they were all laid to rest. We learned more lessons, not the planned lessons, but new lessons about life. We learned that life is fragile and temporary.

People in Houston recognized me wherever I went—at the grocery store, in restaurants, at the airport. They stopped me, encouraged me, and opened their hearts to me with their stories. I learned a mighty lesson from them.

Everyone meets adversity at some time in their lives, they told me. People lose loved ones in death or divorce. They suffer great illness or sorrow. Some of them live quiet lives of desperation, or seek fame only to fail, or seek fortune only to lose. Watch those who suffer a common event in their lives. See how one will become broken and bitter about it, while another will overcome and become greater because of it. The real worth in life, they confessed to me, is loving, learning, and helping others.

What powerful messages! These people had known the depths of despair and the heights of great joy. They had witnessed both the calm and the turbulence of the sea. I mused about my life—the opposites and equals—and with a start I suddenly realized I had not known the joy in life that is equal to, but opposite the sorrow. I wondered if the energy used up in grief and anger could instead be used for a more creative purpose.

Would the resulting joy reach the heights opposite, but equal to the depths I had known? Could it come from real, unselfish *love*? Could it come from learning about and living the true life planned for us? Could it come from helping others—not hovering over them, but real concern—help that really makes a difference?

Technology Versus People

Putting my sorrow aside, opening my eyes to life beyond my front door, I saw a nation's can-do spirit fading. I began again to watch the evening news, ask questions, and read the reports. The study by the Rogers Commission to learn what happened to cause the *Challenger's* fiery explosion had been completed. The news was awful for my family, all seven families, the NASA family, and our nation.

Over-confidence, poor management, neglect, flawed O-rings, poor decisions about the limits of technology—science had jumped ahead of the collective conscience of humanity. Technology was leading us into the future, not people. Something had to happen to shake us, to wake us to this abandon. It did. At terrible expense and unbelievable loss, we learned the mighty lesson that moving forward in our nation's pioneering effort requires people working together as a team, focused not only on where the technology can take us, but also on the education and conscience of the people who are taking us there.

When my husband and his six friends died aboard the fiery explosion of *Challenger*, a piece of our nation's can-do spirit

Art: Bob McCall

died too. The nation's pride in its space program had spilled over into our lives both in the work place and the home. It kept us on the cutting edge of technology and the frontiers of both the mind and space. As a nation, if we could put a man on the moon, we could accomplish most anything, we told ourselves.

The miracle of space travel made the symbolic chest of Uncle Sam swell with pride. Our country's space program had taken us to great heights. It was challenged early on by President John F. Kennedy in a speech he made at Rice Institute on September 12, 1962. His words have even more meaning today: "Surely the opening vistas of space promise high costs and hardships, as well as high reward."

Photo: NASA and Frank Culbertson

They slipped the surly bonds of earth to touch the face of God

The Mission Continues

Pioneer Spirit

With Alan Sheppard's historic flight, America had a glimpse of the "vistas of space." Again, history was in the making with John Glenn's incredible orbit of the earth, and then an orbit of the moon, and finally Neal Armstrong and Buzz Aldrin's steps out upon the moon. What a heroic demonstration of courage for the astronauts who stepped over the edge for humanity! What a powerful demonstration of the pioneering endeavors of our country and its people who believed in learning and reaching and discovery!

America has always been shaped by the spirit of exploration and discovery. Each generation of Americans has journeyed along an "exploration road." Whether that road was on land, in the sea, or in the air, those explorations and discoveries forever changed the face of our nation—and often the world. While this road took us up and out into space, it also brought us back to our planet when that one strikingly beautiful photograph of Planet Earth, taken from the Apollo spacecraft on its way to the moon, forever changed the way we saw ourselves and ushered in the environmental age.

Art: Bob McCall

Art: Bob McCall

We feared the *Challenger* accident might be the final story in this country's history book of space exploration. Would the next generation have no explorers? Would this tragedy cause a dead end on the "exploration road" to space? On the day of the accident, I pleaded for space exploration to continue. It must, for all those who traveled that way before, for our nation to overcome its grief, for our loved ones aboard *Challenger*.

They were on a great pioneering mission to fly, explore, and teach. Christa, our beloved friend and teacher, was a pioneer in her own right. Their mission must continue, I reasoned, or they've all died in vain. America's mission of exploration must continue. *Challenger* would not be the end. It would be a great transition chapter in America's story of pioneers on the space frontier.

The Phoenix

The nation was aching with a loss of its pioneer spirit. Sometimes acceptance is too difficult, and the cost too great. Something must happen to bring a "phoenix" out of the ashes. To overcome this tragedy, the nation and the families who lost the most should join together to create a cause greater than ourselves. That cause would take us into the future and heal our wounded spirit.

The healing had begun. People throughout our nation approached the families to build memorials to our fallen astronauts. Requests and letters and cards were mailed to us, enough to fill a large room seven times over. We wanted to return that outpouring of love. We came together in my home, sitting around the coffee table in my living room, to decide together how we could respond to this expression of love, and also to a nation that had helped us and that we now wanted to help. Although most of the families lived in Houston, others came from California, New Hampshire, and Maryland to meet, talk, and plan.

We created an office in my home. I hired a secretary to help us get organized. The families selected me to be their chairman. We created a foundation. That spring we incorporated as a non-profit organization for education. We called it a living memorial with no walls, no bricks, no mortar. We worked together as a team, responding to requests and creating our plan.

The world knew that seven *Challenger* astronauts died, but they were more than astronauts. They were our families and friends. The world knew how they died; we wanted the world to know how they lived and for what they were willing to risk their lives. So, you see, we couldn't let them die in vain. Their mission became our mission.

Together, we created a mission to continue theirs. We named it the Challenger Center for Space Science Education. On mighty wings it would rise out of the ashes to create a loving and living tribute to the *Challenger* Seven and to all people who still believe in dreams, reaching for stars, and the pioneer spirit. The following letter represents our philosophy and reasoning behind the Center:

Letter to America
from the families of the *Challenger* crew
January 28, 1987

One year ago, we shared a terrible loss with you. The *Challenger* crew were our husbands, wife, brothers, sisters, mother, fathers, daughters, and sons—the fundamental, irreplaceable people in the fabric of our lives. At the same time, they—their mission, their quest, their essence—were an intrinsic part of national life too, part of the great extended family known simply as "Americans." They were our pioneers. Together we mourned them and the shortness of their lives. But, in their short time, they contributed and left memories.

They were not people who cherished the soft and easy life, but people who worked hard to extend the reach of the human race no matter what the sacrifice. They risked their lives, not for the sake of aimless adventure, but for the nation that gave them opportunity, and for the space frontier which was an extension of its spirit. They were scientists, engineers, and teachers, guiding us to space. *Challenger's* mission—to give ordinary Americans access to space, to push scientific discovery forward—was a culmination of their work, a fulfillment of their hopes, and an expression of their essential being.

Since their loss, we have been troubled by the incompleteness of their mission. Lessons were left untaught; scientific and engineering problems were left unsolved. Perhaps saddest of all is the idea that America's children must once again put their dreams and their excitement about the future "on hold." This is too great a loss, one we cannot accept.

We wish to carry on *Challenger's* mission by creating a network of space learning centers all over the United States called, cumulatively, the Challenger Center. We envision places where children, teachers, and citizens alike can touch the future. We see them manipulating equipment, conducting scientific experiments, solving problems, working together—immersing themselves in space-like surroundings and growing accustomed to space technology. As a team, they can practice the precise gestures and the rigorous procedures that will be required of them on the space frontier. They can embrace the vision and grasp the potential of space too.

Though it will take time and money to build, the Challenger Center is our idea of a fitting tribute, a celebration of our loved ones' lives, a triumph over their loss. We hope that by making space-like experiences accessible to all people, especially children, we can prepare them for the day when they will take their own place among the stars.

If they were alive and could speak to all Americans, we believe the *Challenger* crew would say this: Do not fear risk. All exploration, all growth is a calculated risk. Without frontiers, civilizations stagnate. Without challenge, people cannot reach their highest selves. Only if we accept our problems as challenges can today's dreams become tomorrow's realities. Only if we're willing to walk over the edge can we become winners.

The families met frequently to plan and create the Challenger Center. We needed more than a plan, though; we needed a board of advisors, a resourceful team, financial support, and lots of old-fashioned persistence and courage to seek help—even help from leaders in the field of the mass media.

We organized ourselves, established our mission statement, hired a small staff, and collected some seed money. Then I placed a most important call to Vice President Bush. I remembered his message in the immediate hours after the accident when he gave us a small piece of paper with his home phone number and the message, "Call if you need us." So months later we called, not for us, but for our dream—a dream to see the *Challenger* mission completed.

Photo: Challenger Center

June with Walter Cronkite

The Vice President answered that call in December 1986 in the following letter:

Dear June:

The loss of the *Challenger* was a loss for the entire country, but I am watching you and the families of the other crew members do a marvelous thing: rebuild the mission of flight 51-L and carry on with its work. The Center's memorial in honor of the *Challenger* crew members is an inspiration to the children, teachers, and future leaders of this nation. This useful knowledge will have a strong bearing on the future success of the national manned space flight program.

I'm sure you know you have my full support and personal best wishes in this endeavor.

Sincerely,
George Bush

Because Mr. and Mrs. Bush led the way with that early response, others followed. They led not only in word, but also in deed. Their compassion and words of encouragement gave us the courage to create the never-been-done-before Challenger Center.

The Challenge

Vice President Bush took a giant leap of faith on our behalf. He served as our honorary chairman, helped us with advice, made calls and wrote letters on our behalf, and even gave us his personal check. Together, we solicited help from others and asked people for their counsel or for them to volunteer to join our team on its quest to overcome a great national tragedy.

These people who volunteered to help told me they remembered where they were and what they were doing when they heard the awful news about *Challenger* in January 1986, when the space shuttle blew up before their very eyes. They described the streams of white smoke and fragments of the shuttle that filled the cold blue sky and how the disaster made such an impact on their lives. They wanted to help. They wanted to make a difference.

Because these volunteers accepted the challenge and joined us to lead the way, others followed who have seen their time

Photo: Challenger Center

Challenger Seven families and President and Mrs. Bush at Challenger Center's first awards dinner.

and dollars translated into wonderful educational programs that encourage our youth to reach for the stars and work hard to see those dreams come true. Because they cared, the *Challenger* crew's mission continues. These volunteers stood beside us in our grief, and now they stand with us as partners on the frontier of education. We had challenges to accept and lessons to learn.

Photo: White House

June with Barbara Bush: friends together in education

Courage

We asked our team of volunteers if we could continue the educational mission for the *Challenger* crew. We wished to create simulated space flight for children that allowed lessons in science and math, providing opportunities for problem-solving and working together as a team on their own mission. Some volunteers thought not. "It's never been done, would cost too much, take too long," they told us. They could not defer judgment during the creative process, but where they fell to the wayside, others with great can-do spirit came aboard.

Now mind you, volunteers didn't just jump aboard. No, it took great courage to seek them out and sometimes even greater humility to beg them to join our cause, to be a part of the team that builds Challenger Learning Centers and provides opportunities for childhood dreams to come true. The courage came when I learned to trust the power greater than myself to provide direction, when I learned to put aside my pride and ask others for help.

On September 23, 1986, at an elementary school in Washington, D.C., we made our public announcement to the students and the news media about our intentions to create the Challenger Center. The news fanned out across the United States and around the world, news that the surviving families of the seven-member crew of *Challenger* would continue the *Challenger's* educational mission and create a space center for students and their teachers. Headlines the next day read: *Shuttle families plan space hub; Challenger families planning space center; Challenger families accept the Challenge; A New Quest: The Challenger families launch educational project in memory of astronauts.*

The news was carried over the wire for newspaper services and for television and radio audiences. The news media that had invaded our privacy to tell the tragic story was now our friend, telling the positive story.

Many who heard the news sent notes of congratulations to us for creating a living tribute to our loved ones. They commended me for my courage to speak publicly before dozens of reporters and television cameras. Some were surprised by our persistence and resolve to see the dream become reality.

Our plans were made public, but our prayers were private. Our prayers were for courage and for guidance to know what was best for the *Challenger* Seven and our families and for the children around the world who were still waiting for their lessons from space.

Persistence

At times great dark clouds of frustration would block our path, but unexpected glimmers of hope—really answers to prayer—would light our way. One of those shining stars was my former student, Richard Garriott, who had become a young entrepreneur in the computer game business. I asked him if we could create for a classroom of students an opportunity for them to work together in a game-like scenario to solve problems on a simulated space flight. Could each student work at a station, each dependent on the other to complete a mission aboard a mock-up of a space station and mission control? In a youthful, can-do spirited reply, he said, "Hadn't ever been done, but don't know why not."

Other stars were artist friend Bob McCall and his wife, Louise. They helped us think about and sketch a design for our educational center: a mini-science center, an interactive and high-tech center. With Richard's creative ideas and Bob McCall's artistic talent and imagination, we created the concept and design to deliver the dream to help the nation continue the *Challenger* mission.

Far from being easy, we traveled over some challenging roads to create the Challenger Center and programs. Risks were involved, but we knew that the greatest risk was to take no risk. Our enthusiastic staff and my stubborn will would not be swayed when the going got tough. Sheer determination and tenacity would help us create a tribute to the *Challenger* of which they and our team of supporters would be proud. Our persistence and tenacious resolve to find solutions, overcome obstacles, and solve problems, added to our strong conviction to continue the mission and courage to seek help from others, were the keys to our success.

Art: Bob McCall

Original artist's concept of Challenger Center

Photo: Challenger Center

Tuscon I meeting created the educational program design

Teamwork

The team grew after we announced our mission. We received contributions from children who sent their nickels, dimes, and pennies (one sent his tooth fairy money); the elderly who still believed in the great American dream; and leaders in government, business, and education. To build the best, we needed the brightest and most creative people available to help us. Kathy Sullivan, astronaut and friend, helped us convene our first meeting in Tucson, Arizona, in the spring of 1987. We called on the scientific community, leading educators, computer experts, astronauts, mission specialists, and flight directors to help the educational program design. Others joined us too: corporations, foundations, the media, and individuals who could help us with financial support.

Still others called us to offer their early support. Disney called with an invitation to provide a special day of recognition, a parade, and a major celebration at Disneyland. Lee Greenwood volunteered to serve as a board member and to sing his "God Bless the USA" at our first awards dinner. His help continues today. Our circle of friends grew until the ripples (like those created from a stone cast upon a pond) reached across America and around the world, and most importantly the ripples reached into the classrooms of students and teachers still waiting for their lessons from space.

To overcome not only the grief and sorrow for our loved ones' unfulfilled dreams, but for those of a nation as well, this team of volunteers helped us to create more than a memorial. Together, we created Challenger Center, where students and their teachers climb aboard a child-sized space station and fly a simulated space mission into the future. There they are challenged to apply skills in math and science while working together as a team to complete a successful mission.

President Reagan offered support in the following letter and through his proclamation for the National Challenger Center Day.

Dear June:

Ever since that tragic day last January, children everywhere have been waiting for their lessons from space. I warmly congratulate you on your tireless devotion to continuing *Challenger's* mission and to providing these youngsters with the lessons that seemed lost.

With the Challenger Center there will be a brighter future, full of hope for all our children to explore and to learn about the frontier of space.

God bless you.

Sincerely,
Ronald Reagan

Photo: NASA

National Challenger Center Day, 1987
By the President of the United States of America
A Proclamation

Will America continue to lead the world in space
exploration as we move into the twenty-first century?

The *Challenger* crew, lost one year ago on the twenty-
fifth space shuttle mission, dedicated themselves to
America's leadership in space exploration. That leadership depends not only on our courage and
determination, but also on the knowledge, capability,
and inspiration of our students who will be the
researchers and the astronauts of the twenty-first century.

A goal of the Space Shuttle *Challenger* mission was to
bring the study of space science directly and dramati-
cally into the nation's classrooms.

In recognition of the critical need to provide America's
students with access to outstanding space science education and to motivate study and excellence in science,
the families of the *Challenger* crew established a
Challenger Center for Space Science Education. This
Center will honor the memory of the *Challenger* crew
with an ongoing monument to their achievements, to
their courage, and to their dedication to future generations of space explorers.

In commemoration of the brave members of the
Challenger crew, the Congress, by Senate Joint
Resolution 24, has designated January 28, 1987, as

"National Challenger Center Day" and authorized and requested the President to issue a proclamation in observance of this event.

NOW, THEREFORE, I, RONALD REAGAN, President of the United States of America, do hereby proclaim January 28, 1987, as National Challenger Center Day, and I call on the people of the United States to observe this day by remembering the *Challenger* astronauts who died while serving their country and by reflecting upon the important role the Challenger Center will play in honoring their accomplishments and in furthering their goal of strengthening space and science education.

IN WITNESS WHEREOF, I have hereunto set my hand this twenty-eighth day of January, in the year of our Lord nineteen hundred and eighty-seven, and of the Independence of the United States of America the two hundred and eleventh.

Ronald Reagan

President Bush, in his remarks at the Challenger Center National Awards Dinner in 1989, pledged his continuing support.

The mission of the Challenger Center is very close to our hearts and close to our vision of an America that gives its children the best possible education for their challenging future. The mission of Challenger Center is to spark in our young people an interest and a joy in science, a spark that can change their lives and help make American enterprise the envy of the world. We applaud your private efforts.

We are a nation of trail-blazers, risk-takers, and dream-makers. The Challenger Center believes that the future is bright, if we Americans are dedicated to making it so. That dedication is what we see in the remarkable programs of the Challenger Center.

In January 1986, Barbara and I grieved for the *Challenger* families, as did all of America. The courageous *Challenger* astronauts were very much on my mind earlier this year when we gathered on the twentieth anniversary of the moon landing to chart a course for the new century: first, space station *Freedom*—our critical next step in all our space endeavors—then back to the moon, back to the future, and this time back to stay, and then a journey to another planet. In the next century, men and women will travel to Mars.

I said then that our fallen astronauts have taken their place in the heavens so that America can take its place in the stars. Today, we are moved by the continuing faith and commitment that have moved America's space program forward and brought the Challenger Center into being.

We salute you tonight for all that you have done for our common future, and we pledge you our continuing support as you move forward with the vision and the determination to carry out the *Challenger* mission.

The first Challenger Center was built in Space City Houston with the able and lively assistance of yet another star, astronomer and friend Carolyn Sumners at the Houston Museum of Natural Science. With our dedicated and talented staff, astronauts, NASA mission controllers, educators, and museum designers, we created the prototype Challenger Center.

The Centers began to grow in number, slowly at first, then at exponential rates. East to Maryland; south to Florida; north to Ohio, Connecticut, New York, and Canada; and then west to Hawaii, California, and Washington—all across the maps of the USA and Canada we placed Challenger Center logos to represent locations of our centers. By the tenth anniversary of the loss of *Challenger* and the birth of Challenger Center, we will have built thirty centers and reached millions of students and thousands more teachers in our workshops and televised conferences.

Students throughout North America travel to the Challenger Center for a different kind of field trip, a field trip outside the boundaries of our planet. Teachers climb aboard the space station with their students and fly a simulated space mission into the future where they are challenged to apply skills in math and science learned back home in their very own classrooms. In this way, Challenger Learning Centers become a bridge between students in their classroom and their future, offering hope and promise for a brighter future.

Passersby may stop and marvel at the architecture of a new bridge. They see in it the design and cooperative ventures of the architect and skill of the builders. Each cable, anchor, brace, and archway forms a part of the plan for crossing the dark waters beneath. So it is with the love that built the Challenger Center. It is a wonderful bridge that spans the turbulent river of time. It links what students learn in the classroom to what they dream for the future.

Our students are transported to a whole new world where they discover the importance of teamwork. In fact, the teamwork required to complete a mission successfully is so intense that many corporations send their employees to Challenger Centers to enhance teamwork and cooperation in the workplace. They work together as a team of engineers, physicians, scientists, navigators, and communicators to chart flight patterns, assemble an

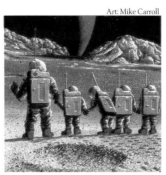

Art: Mike Carroll

The Ultimate Field Trip

electronic probe, solve oxygen and water shortage problems, monitor the health of the crew, and solve problems in a high-tech environment in a true-to-life experience.

On their voyage, the students travel on a scientific mission, work together as a team to solve problems, increase critical thinking abilities, and learn about the fields of science and math. Most of all, they learn about themselves and how to work with others in the world around them.

During the grand opening of the Challenger Center in New York, a teacher entered with her students into the space station simulator. When she saw all the reporters, photographers, and television cameras, she asked Jerome (one of her more animated students) to wait outside. The flight director and I encouraged her to let Jerome join his classmates. She interrupted, "Oh, you don't know Jerome. He'll embarrass us in front of all those cameras. He's been in the courtroom more than the classroom. No, you don't want Jerome." We let her know there was something at Challenger Center for all of her students, including Jerome. Reluctantly, she admitted him, but placed him in the sealed-off and enclosed "probe assembly" room.

Jerome wasn't much of a reader, but he was good with his hands. He experimented with all the mechanical and electronic equipment until he had finally assembled the probe (with some urging from his teammates). When the mission was completed,

Photo: Challenger Center

Students in Challenger Center Space Station Simulator

and the students were congratulating themselves and each other for their success, the media zoomed in on Jerome, who was in his customary state of animation. When one reporter asked him about his duties during the mission, Jerome stepped forward, used the language of the engineer, and took all the credit for the successful mission. His classmates cheered!

By the way, I understand Jerome is still a challenge to work with these days, but for other reasons. His teachers tell me he has recently learned that he is eligible for a scholarship to college and wants to know where the best college of engineering is. The Challenger Center experience is like no other to encourage our youth to reach for the stars and work together to help each other.

More than a classroom of computers and high-tech equipment, Challenger Center is a system that brings community leaders together to benefit their local schools by bringing excitement to learning and providing innovative tools for teaching. Individuals, foundations, corporations, and companies join together in a team effort to create the Challenger Center experience for their schools, science centers, or colleges and universities that want to help take education into the twenty-first-century space age of communication and technology.

The Greatest Lesson

We learned great lessons, not those planned by the "Teacher in Space" mission, but lessons on forgiveness, how to overcome adversity, how to go around obstacles, how to solve problems, seeking God's help, forfeiting control to God, and welcoming new life. How to overcome a tragedy, adversity, or a crisis isn't as simple as how to turn a not-so-good day into one to celebrate, but they are based on the same principle. To make a positive out of a negative, you must first turn the energy of grief and sadness into the energy of creativity.

Life is a series of lessons that help us to grow. One of these lessons is how to overcome a crisis. The Chinese have a word for crisis that explains the process very well. This word consists of two pictured characters: one means trouble and the other opportunity. So when you meet a crisis, tragedy, or trouble, turn the energy used for anger, fear, or grief (that we often bury ourselves in) into energy to solve problems and create an opportunity to bring triumph out of tragedy.

The greatest lesson—the key ingredient, the spark that ignites the creativity, brightens the way, heightens our sense of being alive, strengthens our courage, lessens the burden, puts the magic into our relationships and the sparkle in our lives—is love. I am speaking not only of the love we receive, but of the unconditional love we give—the love that allows us to free others from our control and lets them return love on their own, the love that completes us and connects us with the people with whom we journey through life.

For the *Challenger* families, a whole nation shared in our loss and love, not just the love we returned to an aching nation in turn for the tremedous compassion shown to the families and friends of the *Challenger* crew, but the true, unselfish love and encouragement they gave us to allow us to see this work completed. I experienced the kind and patient love of the families who helped me and put up with my seeming waywardness when many of them would have preferred their privacy. I felt the love of our friends at NASA who helped us when many of them wanted to leave the memory of the accident behind them.

I also learned of love from those who used their creative talents to build the Challenger Center, who responded to my call, who with wrenched arms from the task gave unselfishly of their time and talents. The greatest love was not the love I gave, but the love that was given in faith to see a dream realized—a dream greater than ourselves.

Franklin D. Roosevelt once said, "We cannot always build the future for our youth, but we can build our youth for the future." That's what those great people accomplished when they created Challenger Center. They made a difference. They touched the future not only for Christa and the crew of 51-L who set out to explore, inspire, learn, and teach, but also for generations of children who will reach out to the stars and work hard to see their dreams come true.

The greatest lesson is that God's love reaches into the future. It has no limits and knows no boundaries. It can change the complexion of the universe or the color of our world and turn tribulations into triumphs.

The Power of Love

What lessons have I learned? I've learned, as Matthew Arnold expressed it, "to see life steadily and to see it whole." I've learned that beauty in life and happiness are not found in the controlled, forced environment of my doing, but in the joy of experiencing life freely, naturally, without the clutter of anger, envy, fear, or guilt that stifles and bogs down our lives. It is found in the sheer, sweet pleasure and joy of experiencing life at its fullest and most pleasurable moments.

I've learned that when one door is closed, God opens another. To close the door on yesterday allows the door of today to swing open more fully. Dwelling on what was or might have been steals from us what is today or can be tomorrow. If we dwell on the past, we may miss a quiet moment of solitude with a rocking-chair philosophical acceptance of life that allows God to help our hearts sing again, or we may miss small snitches of time stolen to work the soil of a quiet cottage garden laden with blooms bursting with color in the warm June sun.

We may pass up the opportunity to see the world through the wonder and magic of a child's imagination, sipping make-believe tea from a toy teacup with grandchildren, watching them build a mighty tree-top spaceship, fighting off imaginary pirates. We may fail to experience the joy of intimate conversation that nourishes the soul and fuels the senses like the gentle, burning embers that glow softly from the hearth and warm the winter's night chill. All of these are moments to treasure when the burdens of yesterday give way to the joys of today.

Photo: David Potts

"Amma" June with grandchildren ...

Life—more to live, joys to discover that reach the heights equal to but opposite from the depths of despair. I've been there on that sea, acquainted myself with its many faces, from the calm to the turbulent. I've known its beauty as well as its wrath. I've met a mighty power who helped me overcome adversity, solve problems, and weather the storms of anger, guilt, and self-pity—the storms I battled openly on the sea and those I fought inwardly to test my faith, pride, and humility. Greatest of all, I learned lessons about love, the beautiful power of love that can cause a magnificent phoenix to rise up out of the ashes of a tragedy.

... and "Oma" June's Cristi Lynn

Photo: Houston Chronicle

President and Mrs. Reagan with June at memorial service,
Johnson Space Center January 31, 1986

That love was expressed eloquently by President Reagan when he and Nancy joined the relatives and friends of the *Challenger* crew a few days after the accident at a memorial service at Johnson Space Center in Houston near where we lived. My family sat to their right; Jane Smith's family sat at their left side. We all sat together in silence, some holding small American flags, others clutching handkerchiefs, one cuddling a brown teddy bear, and another cradled on the shoulder of her mother. Music played, the sky was cloudy, and President Reagan rose to speak.

The sacrifice of your loved ones has stirred the soul of our nation, and, through the pain, our hearts have been opened to a profound truth. The future is not free, the story of all human progress is one of struggle against all odds. We learned again that this America was built on heroism and noble sacrifice. It was built by men and women like our seven star voyagers, who answered a call beyond duty.

He paid tribute to each of the crew members, and then added, "Your families and your country mourn your passing. We bid you goodbye, but we will never forget you."

I held as tightly to those words as I did to my composure, until the end of the ceremony when we turned our eyes to the sky to observe an aerial flyover of a missing-man formation of T-38 jets. When the symbolic void was created by one of the jets that represented the death of my husband and our friends, I cried helplessly in spite of cameras and photographers poised to capture my moment of anguish.

Many times during the symbolic flight, I had stood with friends who had lost their husbands, and now they stood with me. I suddenly realized the simple and beautiful truth of the missing-man formation. Though a crew member is missing, those who remain in flight will carry on the mission. As I held my hands to my mouth to muffle the sobs, I looked to the planes and the cloudy sky beyond and made the solemn promise to continue that mission.

The Mission Continues

Does the *Challenger* mission continue? Yes, each day that children arrive at the doors of a Challenger Center, the mission continues. I know because I see it. I see it in the joy on the faces of the children with broad smiles. I hear it in their voices of serious commands, scientific reports, and great cheers and shouts of jubilation. I feel it when I see kids turned on to math and science. I sense it when their teachers tell me the delightful stories about increased self-esteem for underprivileged children or about complex problems solved through a team approach, or when a child sets a new goal and finds a dream to follow.

Most of all, I know the mission continues when I glance to the heavens and see a formation of "laughter-silvered wings" take flight. Sometimes at night I see seven stars twinkle and know seven great people look down on us and smile. When I hear thunderous bursts from the clouds above, I wonder if it isn't applause for the thousands of friends, sponsors, and supporters whose gifts of love have made all the difference. If I use my imagination, I can hear the commander's voice, steady, in control, communicating to mission control: "Throttle up. All systems go." I can hear Christa's enthusiastic voice, rippling through space and time, saying, "Our lesson for today is..."

I hear it at the Challenger Learning Centers all across America and in the classrooms from bustling inner-city schools to quiet, rural towns. The mission continues when a young man or woman turns seriously to ponder where we are in time and space, when a teacher turns from the text to challenge a student to dream big dreams, and when obstacles are met and challenges accepted.

Photo: Chattanooga Free Press

Margaret with "Amma" June blowing out candles on ninth anniversary of the *Challenger* mission and grand opening of 25th Challenger Center at University of Tennessee, Chattanooga

In 1986, we were unified as a nation in disbelief and great sorrow. Today, we join in celebration for this decade of reaching, teaching, and touching the future. Is the mission completed? Well, it has begun. It's moved on beyond our early vision. It belongs now to the future, for generations of teachers and their students to use, improve, and enjoy. Like Coleridge's Ancient Mariner who stopped the wedding guests one by one to tell his story, share his grief, and explain his passion and compassion, I too tell this story, not because I am proud, but because I've learned not to be too proud.

Touching the Future

On April 11, 1995, as we approached our tenth year of continuing the mission, all of the *Challenger* families gathered in Washington, D.C. with our friends and former President George Bush and his wife Barbara to celebrate how far we've come and their leadership and that of others who followed them to fulfill the dream. Steve McAuliffe was there, and so were Marcia Jarvis, Jane Smith, Cheryl McNair, Lorna Onizuka, and Chuck Resnik. I was thankful for their friendship, for our bond that went beyond suffering a tragedy together, a bond of great strength forged from working together on a dream, a mission greater than ourselves.

During the awards ceremony we presented to our two dear friends, President and Mrs. Bush, the Challenger Center Presidential Award (in the future to be called the President Bush Award). They accepted the award and thanked us graciously. In their delightful, light-hearted, and loving style, they gave the following speeches, with Barbara speaking first.

Thank you very much, June and all the members of the extended *Challenger* family, for this special honor; and thank each of you for the great welcome that you have given us tonight.

I can't help but feel that this is backwards; George and I should be sitting in the audience and congratulating you. Will Rogers once said, "We can't all be heroes because somebody has to sit on the curb and clap as they go by." Well, that's what we're doing tonight. We're clapping long and hard for all of you, for everything you have accomplished in ten short years.

Like most Americans, I remember exactly where I was on that fateful day in January 1986: a hotel room in California, sitting and watching television in stunned silence. While you, the members of the *Challenger* family, struggled to deal with your own personal grief, the entire nation was gripped with a great sense of sadness and tragedy.

A Greek philosopher once said, "The test of courage is not to die, but to live." You met that great test with incredible strength and determination. Then you chose to do much more, to bring great meaning to your life and to the lives of those lost aboard the *Challenger* by turning grief into a positive force. You helped us heal the wounds much more than we were able to help you.

June, you truly are one of my heroes. As the founding chairman of the Challenger Center, you worked tirelessly to make this dream possible. At a very difficult time in the history of our country, you were the backbone for us all, and for that, we'll always be grateful.

A few years ago, when George was president, I visited one of the Challenger Centers: the Howard B. Owens Science Center in Greenbelt, Maryland. What a wonderful place! It was full of knowledge, enthusiasm, and fun. Through the Challenger Centers, now opening all over the country, you have given the children of our country a great gift: the gift of hope and the gift of learning.

Much has happened in the space program and in the world in the last ten years–some of it good, some of it bad. But just think: As we stand here tonight, an American astronaut is orbiting the earth with a group of Russian cosmonauts, working together in friendship and peace.

You definitely can take some of the credit for that huge achievement . . . It was your courage and determination that helped us all keep the dream of exploring outer space alive.

Thank you so much for making us a part of this very special evening.

Barbara Bush

Photo: Helen Exum, Chattanooga Free Press

June with President George Bush and Barbara Bush at 1995 Challenger Center Awards Ceremony

After Barbara finished speaking, she passed the microphone to her husband, giving a clever mimic of heralding trumpets to introduce the former president of the United States of America. Mr. Bush spoke these words:

Nine years ago, an inspiring idea blossomed in the midst of great pain and sorrow. The idea was in a special way a powerful expression of a uniquely American trait: the ability to forge triumph from adversity.

The creation of the Challenger Center for Space Science Education and the founding of Challenger Centers across the country was a remarkable demonstration of strength and goodwill and an abundance of faith. Just as remarkable has been the good work that has come out of these centers of hope and learning and inspiration. I think it's wonderful seeing these kids here tonight.

So tonight we celebrate the continuation of a dream and the power of an idea. Through their work, the people of the Challenger Centers have taught thousands of youngsters the values which have, for more than two centuries, made our country strong and vibrant. Those who have been touched by a Challenger Center have learned about teamwork, about challenge and response, about dealing with adversity and learning to solve problems, and perhaps most importantly, about the fundamental lessons of human interaction.

It has been said before, but it certainly bears repeating–especially tonight: the dream is alive. And you have made it grow.

The hopes and dreams of the *Challenger* crew were as varied as their backgrounds. They came from many places and more experiences, but they came together as a team, united by the power of an idea so simple and yet, so powerful: to teach.

Along the way, they hoped to inspire and make real in the hearts of our young people the certainty that dreams do come true, that adventure and exploration and discovery await those who take the time to prepare and have the will to make it happen.

Dick Scobee, Mike Smith, Greg Jarvis, Judy Resnik, Ron McNair, Ellison Onizuka, and the remarkable schoolteacher, Christa McAuliffe, continue to live on in our hearts. We honor their memory tonight; but moreover, the Challenger Centers honor their memory–what they stood for–every day they open their doors.

You've come a long way in those ten years, but challenges lie ahead. As I see it, it boils down to two great challenges. The first is to continue to help the Challenger Centers grow. From its first learning center in Houston in 1988, the learning centers have spread to twenty-five cities. More than 600,000 students, teachers, and parents have been a part of Challenger Center programs; and thousands more have benefited from the teacher training programs. We must continue to nurture these efforts.

The second task is perhaps a more daunting one–but no less important. It is to see that the spirit of exploration and discovery continues to endure in this country. It's to make sure that American space exploration continues to push back the frontier on into the next century.

Of course, it won't be easy. The space program is confronted with its own problems these days . . . adversity, technical problems, the drive to balance the federal budget. . . . As president, I never wavered in my support for the space station *Freedom*. We kept the project fully funded and on schedule, and though there have been setbacks, I hope we won't abandon it now. We also defined the new frontier in space exploration when we committed this nation to going to

Mars. We raised the bar a notch, and we did so with the pride and confidence that we could reach that goal.

Never before in our nation's history has America surveyed a frontier and then retreated. We have always been a people among whose basic urges is to seek out the distant horizon, to surpass it, to go to the next, and the next after that. We are the seekers, the explorers. We are the leaders. It is the stuff of our history–the stuff of our past. I believe it should also have a place in our future.

The *Challenger* Seven lived in vibrant pursuit of a dream. As long as we continue to pursue that dream, as long as we help it to touch the lives of our young people, as long as we help to ensure that America continues to rise to the challenge of the new frontier, then it can be said that we never truly lost those seven brave souls. They will continue to live on with us, and with the hopes and dreams of the nation.

George Bush

We are truly blessed to count these two caring leaders as our personal friends as well as friends to the Challenger Center. They have always been there for us. They have joined us at somber memorial services, at the graveside of our loved ones, in their offices at the White House, in their homes, at receptions to commemorate the loss of the *Challenger* crew, and at celebrations of their continued mission. Since the day he first joined us at the crew quarters at Kennedy Space Center in Florida, President Bush has kept his promise to answer our call when we needed him.

Standing beside him the day of the accident was Senator Jake Garn, who also has spoken in our favor and helped us over many hurdles. He has personally encouraged me with the following letter that I treasure, not only for its contents, but also for the humility it has taught me.

Dear June:

All of us, I suppose, at one point or another in our lives, wonder what we might be able to accomplish that would truly serve as a legacy—what single contribution we could make that helps others for which we'd like to be remembered. Most people probably never fully find the answer to that question or don't have the opportunity to see the results of their efforts come to fruition. You are one who has been blessed with both the knowledge of your purpose and the opportunity to see it become a reality.

As with all great accomplishments, what you have done with the Challenger Center did not just happen. It took an incredible amount of courage, commitment, devotion, dedication, talent, ability, and perseverance on the part of a great many people. And you, as the leader of the *Challenger* families' efforts to more than memorialize your lost loved ones but to institutionalize their mission and their collective dreams, must be given so much of the credit for their success.

I will never forget the feelings you inspired in me as I stood in the crew quarters with Vice President Bush and John Glenn and listened to you as you stood and spoke for all of the families. There you were, only hours after the accident, and in the midst of your own personal tragedy, speaking with such courage and conviction about the need for the space program to go on and for the mission of that flight to continue.

Had you said or done nothing else beyond that moment, I would have counted you among the most unforgettable, inspirational people I have ever met. But that selflessness and courage and the vision you shared carried you beyond that moment, beyond the memories and into the future. It has enabled you to bring to maturity the Challenger Center program, which offers opportunities for young people that will completely change their lives and will change the direction of the nation those young people will one day lead. Because of that, you and the other family members have literally helped shape the destiny of this great nation, and you deserve the highest commendation and the greatest possible applause for what you have accomplished.

I know there is more to be done, and I know you will continue to be actively involved in it. But you are equally deserving of some time for yourself and your family. I am so happy that you have found a companion to love and share the rest of your life with. As you know, I speak from very direct and personal experience, and I know how much more rich and full and meaningful your public life can be when you once again have someone in your personal life to share it with.

Please accept my most sincere congratulations and appreciation for everything you have done. You will always be an inspiration to me.

Sincerely,
Jake Garn
United States Senator

New Beginning

Ten years of touching the future; the Challenger Center; inspiring, exploring, and learning lessons—but the greatest lesson is love. Nothing is more important—not fame, it's fickle; not fortune, for more money never added anything of great significance to a person's life. The real worth in life is loving and learning that the light of love will guide us on our journey, no matter how rough the terrain or how difficult the storm. After the darkest of nights come the dawn and a new tomorrow to offer a new beginning.

My children had learned that lesson well by the second year after their father's death. Rich completed his final semester of college, graduated from the Air Force Academy, went on to pilot training, and then married his high school sweetheart. My daughter Kathie had carved out a beautiful career as a public relations director for colleges and created a lovely home for her family. Both had overcome their grief and moved forward to new beginnings in spite of their mother's inability to overcome the sorrow and loneliness in her life.

June with Rich and Mom and Dad Scobee in front of F-16

Photo: David Potts

Kathie with husband Scott Fulgham, and
children Justin, Emily June, and Courtney

Rich reunites with Cristi Lynn and Dexter
after tour of military duty in Saudi Arabia

Sundays were especially lonely for me those first two
years. After I moved from Houston to Washington D.C., my children and other family members called regularly. Sometimes I
heard from friends: Nancy who cared and made me laugh, Pat who
knew how to picnic even on a rainy day, Diane and Kirby who
encouraged me, and Kathy who was thoughtful and fun to talk
with. Others, such as Elaine and Barbara, wrote meaningful notes.
But they were all miles away in Texas, California, England, and
Germany. I missed them, our conversations, and having dinner
with someone who cared.

I was envious of every elderly couple I saw. Why were some of us more fortunate to grow old with our loved ones? During the weekdays, I lost, even buried, myself in the work of first the university, then Challenger Center. Sundays had always been reserved for family, though, and I didn't know how to change that tradition and had not made the effort to make new friends. Everyone I knew was married, I rationalized, and wanted to be with their own families.

During the second anniversary of the loss of *Challenger*, several reporters interviewed me for news or magazine stories. One of those reporters was Jane Mahaffey, a young widow and talented writer. We were immediate kindred spirits, and both of us looked forward to seeing each other again. Jane invited me to an Easter sunrise service to be held at Arlington National Cemetery, where Dick was buried. Having no other plans, I accepted without hesitation. Others who had been widowed would be there as well, and one man in particular who had just lost his wife only the month before needed friends for support, Jane told me.

That Easter morning, the alarm clock buzzed with loud indignity. Groping in the dark, I switched off the alarm and turned over to sleep again, tugging at the cover for warmth. I'd already decided it was too dark for a single woman to be alone on the streets and way too cold to sit through an outdoor sunrise service, chilled to the bone. I would explain my absence; Jane would understand. I dozed until a lightning bolt of guilt flashed through my conscience and brought me quickly to my feet. Racing to get dressed and out the door, I thanked God for His saving grace and for my new friend whom I didn't want to disappoint.

When I arrived, Jane was there with the others waiting to meet me on this special holiday. Easter that year was on the third of April, a much colder Easter than I had ever experienced in the South where I grew up or in Texas and California where Dick and I had lived for so many years.

The service began. "Easter sunrise services mark the dawn of Easter, Christ's promise of new life for all people," we were told. "It's the most important holy day of the Christian religion." The dark gave way to light. The sun, a rosy-orange ball, first peeked over the horizon and through the Greek architecture of the amphitheater, then turned to a yellow glow and climbed in behind a sky of scattered clouds and spread its rays gently over us where we sat together. Rebirth, I thought, when nature awakens, when the warm sun meets the snow of winter, when spring brings forth the green grass and bouquets of flowers and baskets of colorfully decorated eggs for children.

As I listened to the sermon, I was encouraged. I thanked God for the challenge of yesterday, the beauty of the day, and hope of a new tomorrow. I prayed for a map to point the way to a more hopeful future. I wanted the true light to shine within me until all the darkness of fear and loneliness had subsided. I wanted to let go of the vulnerable self within and reach out to the people weeping softly beside me and to others I knew, but the energy I used for my own sorrow sapped from me the strength to help the others.

The chaplain grew silent, and then with a strong, determined voice said, ". . . To have new life, new friends, new experiences, we must let go of the pain and close the door to the past. We must learn to trust our higher power to lead us safely through to the new day." The thought was comforting and gave me courage to reach out to touch the others, to smile and nod my understanding to them.

After the service, I walked across the street from the amphitheater to Dick's gravesite, wished him a happy Easter, and returned to the others to say goodbye. The day was young, and since none of us had family to return to, Jane asked if we would all join her for a walk along the Potomac River.

Mild chaos broke out among the group in trying to decide the logistics of changing into casual clothes and shoes and managing all the cars and directions to the designated meeting place. I waited and listened to the different ideas, wondering if anybody would take charge. I needn't have worried though, because the general decked out in his ribbons and three stars calmly stated the plan and location and clear directions. For the first time, I really looked at this man. He was tall and slender and had piercing eyes and a gentle voice. I saw a man of admirable strength in spite of the recent loss of his wife. Captivated by his actions, he had my full attention and the attention of the others.

Later that day we walked together along a well-worn towpath beside the river, over large rocks, under branches, to the cliff's edge, through a meadow until the others ran ahead and left me walking alone with the general. We talked more openly with each other then. Each understood the other's loss and pain and concerns for the future. The sun shone more brightly. Puffs of clouds in the cobalt blue sky reflected in the river and eventually vaporized into a pale blue sky. The warmth of the day and our conversation brought a peacefulness over us like a gentle rainshower brings freshness to the spring day.

This new friend's name was Don Rodgers. He had lost his wife, Faye, to a sudden heart attack as she was driving home on a crowded freeway in Washington, D.C. A good Samaritan pulled off the highway and parked behind her car to help her. He called for an ambulance when he saw her slumped across the steering wheel, and then he called Don's office at the Pentagon. Finally, a policeman came for Don to rush him to his dying wife in the emergency room of an Alexandria, Virginia hospital.

Seeing his wife's lifeless body attached to tubes, a breathing apparatus, and electronic devices was traumatic for him. Her doctor gave no hope or encouragement. Don called for their only son Eric, who lived in Dallas, to fly to Washington to his mother's bedside. In only a few hours, Eric was there embraced with his father, holding onto his mother until all life left her, leaving behind the family she treasured.

Walking alongside each other, remembering and talking but mostly listening, we understood the other's loss and appreciated each other more, knowing the sorrow that filled our hearts. Don's loss was not public like mine, I thought, but just as tragic for those who loved her. We agreed that at first we cry out in anguish for our dying loved ones, and then for ourselves—for the great gaping wound it leaves in us that only time can heal. Our conversation was a comfort for both of us.

We returned to the cars where the others were waiting. We thanked each other for the full day of companionship, then parted. Don was being transferred to Arizona to command Fort Huachuca, near Tucson. We wished him well. He turned to me and said, "Thank you, June." He drove off. I never expected to see any of them again, except for Jane who would meet me on Sunday for church services and sometimes lunch. We never know the full impact of a single event in our lives, but I was certain that day would influence all of my tomorrows.

Happiness

The freshness of spring gave way to the heat of summer, and work continued to fill my days. I criss-crossed the country in travels as I sought support for Challenger Center and made trips to see my children. By August, Rich had an assignment to Germany to fly F-16 fighter jets at Ramstein Air Force Base. It was difficult to tell him goodbye. I was happy for him in his new assignment, but I would miss him immensely.

Old habits are sometimes difficult to overcome. To fight back the tears of loneliness and the fear of isolation, I immersed myself even more into my work. I slipped back into my old way of control, relying not on God's light to lead the way, but on my own selfish need to try to control every situation and outcome. When a movie was created for television that neither I nor the other *Challenger* families approved, the stress and anxiety of trying to manage the situation and prevent its release became so overwhelming that I needed medical attention.

With no neighbors to call and my children in other parts of the world, I remembered my new friend Jane and asked her to help me. As we drove to the hospital, I prayed for guidance, for strength to let go, for wisdom to overcome the obstacles in my path. Jane helped with her presence. Someone was there who cared. She listened and cheered me with jokes. Laughter is a great healer, especially when you can laugh at yourself. She helped me to create a humorous image of myself standing atop the world trying to save everybody else, when the person who really needed saving was me. She invited me to talk to our chaplain at the Army post and to a psychologist.

All of these persons helped me to find the answers to my prayers. They led me to see that no one is perfect, that feigning great strength is indeed a weakness, and my need to hold on to the past (to hold on to Dick Scobee) was a selfish act, not only for those who loved me but for myself. For my own health and well-being, I learned that I must first turn away from grief, turn away from my loss, and turn toward God and the gift of new life to discover new friends and happiness.

Guided by the light of love and laughter that healed and softened the jagged edges in my life, I soon had the strength and courage to begin letting go of the need to hold on to my past and control others, and to begin to create time and opportunities to make new friends. After church the next Sunday, I drove to the cemetery where Dick was buried. I knelt at his grave and prayed for strength to move forward in my life, to accept my life as it was, to put God back in control. I needed to hold on to the memories, but to let go of the past, to let Dick Scobee return to a higher plane, to his star that would glow in the heavens forever.

Several weeks later, the general called, or I should say, his military aide Brian called for him. Don would be in town and wanted to see me. He had heard that the nice lady he met at Easter had been in the hospital. I agreed to meet him when his plane arrived at Washington National Airport. As I waited at the airport for the man I had met at Easter and seen briefly at dinner in August, I wondered if I would even recognize him. I looked around to see if there were other women my size waiting alone. Relief! There were none, so even if I didn't know him, hopefully he would recognize me.

One flight of passengers arrived, then another. Turning the corner and coming into view was a familiar face, but it wasn't the tall, distinguished general I remembered. No, this man wore a smile across a grey and ashen complexion. Looking unusually thin in his slim cut jeans, he walked toward me. Sadness filled his eyes.

He was alone, no military aides or friends around him. Dressed in my pale pink suit, I reached out to hug him.

Standing there together, we must have looked ghostly: two lonely people, survivors whose paths had each ended abruptly with the one we loved, had taken a turn on different journeys, and now had come together. We each needed a friend—someone who cared, someone who would call, someone who would understand.

Months later we admitted to each other that we both knew then that our friendship would grow to love and possibly marriage. Our understanding of the other's loss brought us together in a bond of common need and love, each helping the other, which in turn helped ourselves.

Our friendship grew. We talked on the telephone like teenagers—getting acquainted and learning about each other's families, interests, Christian faith, joy in each other, and hopes for the future. As we healed, we learned to laugh more and to appreciate each other. Our paths had met and stretched out before us as one. I was forty-six years old, and Don was fifty-four. He and Faye had lived a beautiful life together for thirty-two years. Dick and I were married more than twenty-six years. Could we possibly begin a new life together? We wondered, but not for long.

Photo: US ARMY

Lieutenant General Don Rodgers

The tables turned as my children seemingly took on the role of my parents. Don, the perfect gentleman, asked my daughter if he could call on me. Kathie giggled at the thought of her mother dating. He asked my son if he could marry me. Rich answered, "Well, that's up to my mom." When Rich and I talked later, with a grin on his face and twinkle in his eye, we laughed and embraced each other's new joy, new beginnings.

Our wedding date was set for June 3, 1989. We invited our family and friends to the small military chapel next to Arlington National Cemetery where both Dick and Faye had memorial services. What a glorious new day! Standing at the front entrance with my son, who would soon walk me down the aisle, I looked longingly toward the man whose arms I would walk into for the remainder of our lives. He was truly a knight in shining armor, dressed in full uniform. Standing beside him as best man was his son, Eric. To the other side were Kathie, my matron of honor, and her little four-year-old Justin holding a satin pillow with golden rings.

On the right side of the church were seated Don's family and his friends from his military career. In the front pew on the

Photo: Harold Brown

Chapel at Ft. Myer Army Post, Virginia

Photo: Harold Brown

June with Steve and his children Scott and Caroline,
Lorna, Don, Marcia, Jane, Cheryl

Photo: Harold Brown

Challenger Seven families at June and Don's wedding

left were all seven of the families representing the *Challenger* crew:
Dick Scobee's parents, Marcia Jarvis, Chuck Resnik, Cheryl
McNair, Jane Smith, Lorna Onizuka, and Steve McAuliffe and his
two children, Scott and Caroline. They were all there, plus friends
from Houston, the Challenger Center, and the White House.

The soft organ music stopped for a second, and then the
wedding march sounded out like trumpets heralding a magnificent, triumphant event. My son took my arm lovingly, smiled, and
walked with me down the aisle to a new life—two lives, two families, two sets of friends coming together to walk a new journey.

Chaplain Art Jensen led the wedding ceremony. Jane
Mahaffey read 1 Corinthians 13 from the family Bible. As we said

Photo: Harold Brown

Chaplain Art Jensen with Don and June

our vows, our two sons took the candles representing the Scobee and Rodgers families and together lit the single candle that represented our unity. We all came together united in spirit that day just as we all will someday again when we join Dick Scobee and Faye Rodgers, who we knew were together smiling down on us.

That day I discovered joy as great as my sorrow had been deep. God blessed me with another chance to love and be loved, but more important was my rekindled spirit. I learned that the true key to happiness is when we unlock our hearts to give and receive compassion. I had learned to smile within and without and to turn from the past and toward the future.

Survivor to Dream-Maker

When I count my many blessings, at the top of my list after my family and friends is my home located in a rural area on the side of a mountain. I love to lose myself in thought while gazing out my window from the back porch. Standing there I can see out across the meadow and down to the winding Tennessee River that flows through the valley toward the mountains of green forests that touch a blue sky dotted with puffs of cream-white clouds. I am humbled not only by this beauty, but also by the new life I've been given—an altogether different journey to travel on through a renewed life.

Tears of joy filled my eyes, because I've learned that my narrow picture of what is good and right in my life is not adequate. As victim, the journey was a long, hard, uphill climb alone. To surrender, to give up and give in, would have been easy; but to survive, to accept a new life and travel on a new journey, would have been more difficult had I not learned to trust in God's divine plan.

With God's patience, I learned that the lessons were not there to defeat me, but to help me grow—to move me beyond the old habits of holding on to loss, seeking perfection, and keeping up appearances unnecessarily. I learned lessons about adversity—that sorrow is balanced with joy, that love can bring triumph out of tragedy, that faith can heal a broken spirit and give new meaning to life, and that dreams can come true when we accept our problems for the challenges they really are.

Even more important than being survivors, I think God wants us to be winners. Only as we trust in the power that is greater than ourselves to provide the direction and give our lives new meaning can we see the path that leads us to the destination that He has planned for us. Only through a closer walk with our savior Jesus Christ can we have the courage to boldly walk along the path that teaches us the series of lessons we need in the school of life. When we walk that path and learn those lessons, we rise above our personal needs and turn to God for the inspiration that allows us to be dream-makers, to create opportunities for others that will help them along their journey to discover the silver linings beyond their clouds.

Photo: Patrick Murphy-Racey

The Mission Continues: June visiting one of the many Challenger Centers